Similarity and Clustering in Chemical Information Systems

CHEMOMETRICS SERIES

Series Editor: **Dr. D. Bawden**
Pfizer Central Research, Sandwich, Kent, England

1. Distance Geometry and Conformational Calculations
   G. M. Crippen

2. Clustering of Large Data Sets
   Jure Zupan

3. Multivariate Data Analysis in Industrial Practice
   Paul J. Lewi

4. Correlation Analysis of Organic Reactivity: With particular reference to multiple regression
   John Shorter

5. Information Theoretic Indices for Characterization of Chemical Structures
   Danail Bonchev

6. Logical and Combinatorial Algorithms for Drug Design
   V. E. Golender *and* **A. B. Rozenblit**

7. Minimum Steric Difference
   The MTD method for QSAR studies
   Z. Simon, A. Chiriac, S. Holban, D. Ciubotaru *and* **G. I. Mihalas**

8. Analytical Measurement and Information
   Advances in the information theoretic approach to chemical analyses
   K. Eckschlager *and* **V. Štěpánek**

9. Molecular Connectivity in Structure-Activity Analysis
   Lemont B. Kier *and* **Lowell H. Hall**

10. Potential Pattern Recognition in Chemical and Medical Decision Making
    D. Coomans *and* **I. Broeckaert**

11. Chemical Pattern Recognition
    O. Štrouf

12. Similarity and Clustering in Chemical Information Systems
    Peter Willett

Similarity and Clustering in Chemical Information Systems

Peter Willett

Department of Information Studies
University of Sheffield, England

RESEARCH STUDIES PRESS LTD
Letchworth, Hertfordshire, England

JOHN WILEY & SONS INC.
New York · Chichester · Toronto · Brisbane · Singapore

RESEARCH STUDIES PRESS LTD
58B Station Road, Letchworth, Herts. SG6 3BE, England

Marketing and Distribution
Australia, New Zealand, South-east Asia:
Jacaranda-Wiley Ltd, Jacaranda Press
JOHN WILEY & SONS INC.
GPO Box 859, Brisbane, Queensland 4001, Australia

Canada:
JOHN WILEY & SONS CANADA LIMITED
22 Worcester Road, Rexdale, Ontario, Canada

Europe, Africa:
JOHN WILEY & SONS LIMITED
Baffins Lane, Chichester, West Sussex, England

North and South America and the rest of the world:
JOHN WILEY & SONS INC.
605 Third Avenue, New York, NY 10158, USA

Library of Congress Cataloging-in-Publication Data

Willett, Peter, 1953–
Similarity and clustering in chemical information systems.

(Chemometrics series; 12)
Bibliography: p.
Includes indexes.
1. Chemical structure—Data processing. 2. Information storage and retrieval systems—Chemical structure.
3. Cluster analysis. I. Title.
III. Series
QD471.W578 1987 025′.065412′2 86-31433
ISBN 0 86380 050 5
ISBN 0 471 91463 0 (Wiley)

British Library Cataloguing in Publication Data

Willett, Peter, 1953–
Similarity and clustering in chemical information systems.—(Chemometrics series; 12)
1. Chemistry, Analytic—Statistical methods 2. Cluster analysis
I. Title II. Series
543′.001′51953 QD75.4.S8

ISBN 0 86380 050 5
ISBN 0 471 91463 0 Wiley

ISBN 0 86380 050 5 (Research Studies Press Ltd)
ISBN 0 471 91463 0 (John Wiley & Sons Inc)

Printed in Great Britain by Galliard (Printers) Ltd, Great Yarmouth

For Marie and Barbara

Preface

The advent of computers has brought about a revolution in the ways in which information of all kinds is stored and retrieved. Some of the most striking developments have appeared in computerised chemical information systems, such as are run by organisations concerned with the storage, retrieval and processing of information pertaining to chemical structures. This book discusses computational techniques which may further increase the effectiveness of such information systems; specifically, searching and clustering methods are described which are based upon the calculation of measures of similarity between chemical structures in machine-readable files.

The first chapter of the book presents a broad over-view of the current state of development of computerised chemical information systems. After a discussion of the means by which molecules may be represented in machine-readable form, the chapter concentrates upon two major application areas, these being chemical structure and substructure searching and the study of structure-activity relationships, since it is in these areas that similarity and clustering techniques may most usefully be applied. Chapter 2 reviews these two classes of technique. The discussion includes the measurement of similarity between objects, and the wide range of clustering methods which have been suggested for the grouping of data. Clustering methodologies have been developed and used in a very wide range of application areas, and an attempt is made to summarise what is already a widely disparate literature. Particular attention is paid to criteria which can be used to evaluate the merits of different clustering techniques.

The next three chapters form the heart of the book, and present the results of an extended investigation into the application of similarity and clustering techniques to structure-based chemical information systems. Chapter 3 compares different approaches to the calculation of intermolecular structural similarity, and describes the use of similarity measures to provide powerful new retrieval mechanisms for chemical structure and substructure searching. Chapter 4 follows a similar pattern with the first part devoted to a systematic comparison of a wide range of different types of clustering method; this is followed by a description of the use of one such method in an operational pharmaceutical information system. Clustering methods were first developed for use in the life sciences where a data set may contain only a few tens of objects, this implying that even the most computationally demanding clustering algorithms may be implemented with little difficulty. This is certainly not the case with chemical structure files, which may contain thousands or tens of thousands of molecules, and it is accordingly necessary to devise exceedingly efficient implementations if the methods are to be applied in a chemical context. This problem is discussed in Chapter 5, where algorithms are presented in sufficient detail to allow readers of this book to implement the methods for themselves.

A summary of the major results of the previous chapters and suggestions for further work are presented in the closing chapter.

Many people have helped in the work which led to the writing of this book. I owe a deep debt of gratitude to Prof. Michael Lynch and Dr. George Vladutz for introducing me to the study of chemical information systems, and for encouraging my researches for some ten years. Much of the work described here has been carried out by masters and doctoral research students and research associates under my supervision and I would like to thank in this context Abdelmoula El-Hamdouchi, Alan Griffiths, Hazel Morton, Shirley Perry, Lesley Robinson, Ian Rowlands and, in particular, Vivienne Winterman. Funding for their endeavours has come from a variety of sources including the British Council, the British Library Research and Development Department, the Department of Education and Science, Pfizer Central Research, and the Science and Engineering Research Council. I am also pleased to acknowledge the assistance of the staff of the Computing Services Department, University of Sheffield, for providing the computational resources for the work reported here. Dr. David Bawden has been a strong advocate of the utility of my research work, and encouraged me in the writing of this book in his role as the editor of the Research Studies Series.

Peter Willett, July 1986.

Contents

CHAPTER 1
Chemical Information Systems

1.1 INTRODUCTION

Over the years, many schemes have been developed to describe and categorise chemical compounds (Rouvray, 1977). By the start of this century, the two-dimensional chemical structure diagram had established itself as the prime means of communication between chemists, and it is thus hardly surprising that it has formed the basis for the computerised systems that deal with the storage and retrieval of chemical information. This mode of representation is very closely related to the actual nature of chemical compounds, as determined by the wave equations that describe them, and there are thus few of the complexities and ambiguities that characterise, for example, the complex representations that are required for the computer processing of natural language. In addition, it has proved possible to map the chemical structure diagram onto the hardware and software components of modern day computer systems to provide a user-friendly, graphically-based man-machine interface. Given these factors, it is not surprising that information systems in chemistry are probably better developed than those in any other major discipline.

This chapter provides an overview of the facilities which are available in current computerised systems for the storage and processing of chemical structure information. The material is intended to provide the necessary background for an understanding of the similarity and clustering techniques that form the heart of this book. Accordingly, particular attention is paid to the retrieval of chemical structures and to the correlation of molecular properties with structure since it is in these areas that similarity and clustering methods would seem to be most immediately applicable. More detailed treatments of the topics considered in this chapter are presented in the review volumes of Ash and Hyde (1975) and Ash et al. (1985).

1.2 THE REPRESENTATION OF CHEMICAL STRUCTURES

Four types of molecular representation have been used extensively in chemical information systems, these being systematic nomenclatures, fragmentation codes, line notations and connection tables. Of these, systematic nomenclatures, specifically those devised by Chemical Abstracts Service and by the International Union of Pure and Applied Chemistry, are primarily used in manual information retrieval systems, such as printed indexes, and to support database creation activities (Vander Stouw et al., 1976); however, the lack of flexibility in the representation means that nomenclature often needs to be translated automatically into another type of representation if it is to be of practical use in computerised chemical information systems (Vander Stouw et al., 1974; Willett, 1980). The connection table has proved to be by far the most flexible and generally useful representation, and it forms the basis for most present-day systems, as well as for the experimental work that is described in Chapters 3 to 5 of this monograph. Such systems have been designed primarily for use with two-dimensional (2-D) representations of completely specified molecules, i.e., the classical structure diagram, and the contents of the chapters reflect this bias. Rather different approaches are required for the handling of compounds in chemical patent claims and for three-dimensional (3-D) chemical structures: these topics are mentioned in Sections 1.2.1 and 1.4.4 respectively.

1.2.1 Fragmentation codes

Fragmentation codes were the first structural representation to be widely used for chemical retrieval applications and remain in use to this day, albeit in a very different form, as a component of systems based upon other types of representation.

A fragmentation code comprises a set of pre-defined substructural attributes, the presence or absence of which is used to characterise each of the molecules in a file for retrieval, or other, purposes. It can thus be regarded as a pre-coordinate indexing system, and its degree of success will depend in large part upon the discriminatory abilities of the substructures chosen for inclusion in the indexing vocabulary, and upon the extent to which they reflect the overall structural characteristics of the file of compounds that is to be indexed.

Early fragmentation codes were developed on the basis of subjective impressions of chemical importance, and the attributes which were chosen reflected the typical pre-occupations of the organisation responsible for the creation of the file of compounds. Thus a retrieval system for IR spectroscopy data would adopt a very different fragmentation code from a system designed for the retrieval of organo-phosphorus compounds. The general approach involved the identification of structural features that occurred frequently throughout a collection so that these could be described in some detail, while less common features might be assigned less discriminating notational codes: thus a file containing many potential antibiotics might have several codes for the presence of particular types of penicillin or cephalosporin rings. As we shall see, similar considerations are reflected in modern approaches to the selection of retrieval keys.

A fragment code representation has several limitations. The coding of new compounds as they are added to the file needs to be done manually, and all of the molecules need to be reprocessed if the code is changed in any way (Craig and Ebert, 1969). Furthermore, the use of a fragmentation code results in an ambiguous molecular representation. The representation is ambiguous in that the set of codes assigned to a molecule does not serve to characterise the structure fully, as the fragments might be inter-connected in very many ways, or might not represent the complete set of structural features present in the compound. Accordingly, a set of codes serves to characterise, but not necessarily delimit, an incompletely specified class of compounds, rather than characterising an individual molecule. This feature has several advantages when the generic structures that characterise chemical patents are considered, and fragment codes are hence still used extensively in patent information services. A single such generic, i.e. partially specified, structure in a patent may well correspond to many thousands, or even an infinite number, of individual specific molecules, and this presents severe problems to conventional chemical structure handling systems. However, the great importance of patent information in chemical information (Ash et al., 1985) is now leading to a proliferation of research into such problems (Barnard, 1984), and substantial progress in this field may be expected within the not-too-distant future.

Fragmentation codes were developed originally for use with punched card sorters or optical coincidence cards. However, the advent of linear notations and connection tables, which are described below and which are unambiguous, machine-readable representations, led to a rapid decrease in the use of fragmentation codes as the primary means of structural description. Instead, they are now used for the initial screening stage of substructure searching, as described in Section 1.3.2 and as molecular descriptors for the structure-property correlation studies that are described in Section 1.4.3.

1.2.2 Linear notations

A linear notation is a coding mechanism which allows a chemical structure to be represented by a string of alphanumeric characters: notations are thus related in many ways to systematic chemical nomenclatures, although the latter are of less practical use owing to the implicit representation of many of the important chemical features in a molecule. A large number of notational schemes have been described, and are still being described, in the literature, but only one of these has been used extensively within computerised chemical information systems (Rush, 1976). This is the Wiswesser Line-formula Notation (WLN) (Smith and Baker, 1975) which has formed the basis for many in-house chemical information systems over the last two decades. An example of such a system is the well-known CROSSBOW package (Ash and Hyde, 1975), a set of software routines for the maintenance and searching of in-house WLN files which was originally developed at ICI Pharmaceuticals Division, but which has now been implemented in many other organisations. Examples of public WLN-based services are the Index Chemicus Registry System (Garfield et al., 1970) and the Chemical Substructure Index (Granito and Rosenberg, 1971) produced by the Institute for Scientific Information.

Compounds are encoded in WLN using a symbol set consisting of the upper-case alphabetic characters, the digits '0' to '9', and the symbols ' ', i.e. a space, '&', '/' and '-'. An individual symbol may be used to represent either an individual atom and the bonds surrounding it, or a group of two or more atoms: thus the symbol 'X' denotes a carbon atom attached to four non-hydrogen atoms whereas the symbol 'V' denotes the pair of atoms that comprise a carbonyl group. The inter-connections of the atoms and groups are not stated explicitly, but are implicit in the ordering of the symbols within the string. The meaning of the symbols may be context-dependent, and thus quite detailed analysis may be necessary to determine the exact form of the structure that corresponds to some particular WLN. As an example, a digit such as 5 may denote either an acyclic chain of five carbon atoms or a ring containing five atoms, inter alia, the exact meaning being determined by other symbols in the string: for example, the presence of a 'U' adjacent to an 'acyclic 5' denotes a carbon-carbon double bond at one end of the carbon chain, while the letter 'T' preceding a 'ring 5' denotes that the ring is a heterocycle of some kind. With a modicum of practice it is relatively easy for chemists to convert many WLNs into structure diagrams, but a substantial knowledge of the coding rules is required if the reverse transformation, from a structure diagram to a canonical

WLN, is to be performed. Facilities are available for the description of steroechemistry, charged species and isotopes; a full description of the coding rules for WLN is given by Smith and Baker (1975).

With one or two minor exceptions, where molecules contain elements that may exhibit variable valencies, a WLN is an unambiguous representation since it defines fully the constitution and connectivity within a compound. Moreover, if the encoding rules are followed exactly, the representation is <u>unique</u> since there will be only one possible symbol string for any given molecule. The canonical nature of a WLN makes it very appropriate for compound registration, as described in Section 1.3.1, and for the production of printed inventories and indexes of compounds. Increasingly, however, WLN-based systems are being converted to connection tables: the conversion is normally done automatically with WLN-to-connection table conversion programs typically achieving 90-95% success rates, and with only a small number of connection tables needing to be converted manually (Granito et al., 1972; Ash et al., 1985).

1.2.3 Connection tables

Connection tables now form the primary means of representation for the structures in both public and in-house chemical information systems. A connection table contains a list of all of the non-hydrogen atoms within the structure, together with bond information that describes the exact manner in which the individual atoms are connected together. A table for the compound

$$^{1}CH_3-^{2}\underset{\displaystyle |\atop ^{3}OH}{CH}-^{4}CH_2-^{5}CH=^{6}NH$$

is as follows:

Atom Number	Atom Type	Connections					
1	C	2	1				
2	C	1	1	3	1	4	1
3	O	2	1				
4	C	2	1	5	1		
5	C	4	1	6	2		
6	N	5	2				

In this table, the first line shows that atom number 1, in some arbitrary numbering scheme, is a carbon atom linked to one other atom, number 2, by a bond of order 1. Similarly, atom number 2 is connected by single bonds to atoms numbers 1,3 and 4 etc. Hydrogen atoms are not included since their presence or absence can be deduced from the bond orders and atomic types. The table shown above is said to be redundant since each bond appears twice within the table, once in the linkage lists of each of the two atoms that the bond joins together. Non-redundant connection tables may also be used, in which each such linkage is listed once only. A further type of connection table is represented by the bond-implicit connection tables used in the CROSSBOW system, where the bonds are not defined explicitly but are instead subsumed in the descriptors that are used to characterise each of the atoms in a molecule (Ash and Hyde, 1975). All types of connection table contain much the same information, although this may be described in different ways; for example, there are considerable discrepancies in the degree to which aromaticity and tautomerism are represented. It is, however, usually possible to convert between different types of connection table without undue difficulty.

Connection tables are unambiguous since they give a complete description of the topology of the compound that they represent. However, they are generally not unique since for a compound containing N non-hydrogen atoms, there are no less than N! different possible numbering schemes that may be used to number the atoms. Thus if a canonical connection table is required, algorithmic means must be employed to obtain a unique representation; this point is discussed further in Section 1.3.1.

Given a redundant connection table, it is relatively simple to generate fragment substructures from it so as to obtain an algorithmic fragmentation code. The fragments generated may correspond to chemically significant substructures comparable to those that formed the basis for the manually assigned fragmentation codes: thus Granito et al. (1972) describe the generation of Ring Code fragment descriptors from connection tables which are in turn generated from WLN. Alternatively, and more commonly, the fragments generated will correspond to formally-defined structural features, the generation routines being based upon a systematic consideration of each of the features of some specified type, or types, of substructure that are present in the molecule. Algorithmic fragmentation is discussed further in Section 1.3.2.

Since connection tables contain an explicit representation of the inter-connections between the atoms in a molecule, they are particularly well suited to manipulations involving such topological information; examples include atom-by-atom searching, graphical structure input and display, structure-property correlation and reaction indexing.

1.3 SEARCHING FILES OF CHEMICAL STRUCTURES

Chemical information systems currently offer two principal types of search facility. Structure search involves the search of a file of compounds for the presence or absence of a specified query compound. Such a search is required when there is a need to retrieve the data, e.g. biological or spectroscopic results, which have been collected for some compound, and when a new molecule has been synthesised or otherwise acquired. This latter process of registration needs to be carried out to determine whether the current molecule has in fact been identified previously, or whether it is novel, in which case it must be added to the structure file. Substructure search involves the retrieval of groups of molecules, all of which contain some specified query substructure, irrespective of what other structural features may be present. Both types of search facility are described in this section so as to provide a context for the novel, similarity-based retrieval mechanisms that are presented in Chapter 3. The description is restricted to search systems that are based upon connection table representations of molecular structure; retrieval mechanisms for other types of representation are reviewed by Ash et al. (1985). The reader should note that chemical structure searching using connection tables is a specific example of the more general graph isomorphism problem (Read and Corneil, 1977; Tarjan, 1977) which arises in the area of mathematics known as graph theory. Graph theory considers the manner in which points, known as nodes, are connected together by edges, and the properties of such graphs; in the chemical context, the nodes are represented by atoms and the edges are represented by the bonds that link the atoms together. Because of this analogy, many of the algorithms which have been developed for the manipulation of general graphs are applicable to operations involving chemical graphs. Of particular interest are graph and subgraph isomorphism algorithms, which may be used for structure and substructure searching respectively, and maximal common subgraph algorithms, which are finding increasing application in automatic chemical reaction indexing systems.

1.3.1 Structure searching

Structure searching can take place only if the structure representation is unambiguous, and it will be effected most easily if the representation is also unique, since a character-by-character match will then suffice to compare two structures for identity. If a non-unique representation is used, conversely, a complex series of comparison operations may be needed to determine the equivalence of variant representations of the same structure.

A very simple way of establishing equivalence between a query structure and a structure in the file might be to generate all possible numberings of the atoms in the query, and then to compare each such numbered connection table with that for the structure in the file to determine whether the two connection tables are identical. The difficulty with such an approach is the sheer number of differently numbered connection tables that are generated. For a molecule containing N atoms, no less than N! different tables may need to be considered; thus the compound and connection table shown in Section 1.2.3 above contains only 6 non-hydrogen atoms but would still result in no fewer than 720 different possible numberings of the atoms. Accordingly, the approach is infeasible for all but the most trivially-sized structures. Various strategies may be used to limit the number of possible numberings that are considered, but even then the computational demands are still high.

A simple and efficient match will be obtained if, and only if, one of the many possible tables can be selected as the basis for the representation in the file and in the query, and there has thus been considerable interest in transforming connection tables into a unique, or canonical, form for the purposes of storage and retrieval. The best-known scheme is the Morgan algorithm (Morgan, 1965) which defines a simple and elegant method for producing a unique numbering of the set of atoms in a connection table. A development of this is the SEMA-name, which takes account of stereochemical information as well as the topological data that forms the basis for most connection table systems (Petrarca et al., 1967; Wipke and Dyott, 1974).

An alternative, and highly efficient, means of implementing structure search is to use a hashing function (Maurer and Lewis, 1975) to calculate some code from a connection table. The code for the query structure may then be matched against the hash codes corresponding to each of the structures in the file: a detailed, exact match operation is carried out only for those very few molecules that have a hash code identical to that for the query compound. A useful feature of methods based upon hash codes is that they may not require the generation of a unique connection table: the systems used at Pfizer Central Research and at Chemical Abstracts Service are described by Bawden et al. (1981) and by Freeland et al.

(1979) respectively.

1.3.2 Substructure searching

Substructure searching involves the retrieval from a machine-readable file of all molecules which contain some specified query substructure. Thus a chemist might be interested in compounds containing the -CH=CH-NH- grouping, in steroid rings which are substituted by at least three hydroxyl groups, in poly-unsaturated alkyl esters etc; in all cases, the molecules retrieved contain the specified features, irrespective of the surrounding structural environment. Such searches are inherently more complex than registration, and belong to the class of computational problems which are known to be NP-complete and thus not susceptible to efficient polynomial time algorithms (Tarjan, 1977). The complexity of substructure searching arises from the fact that the canonicalisation and hashing procedures described in the previous section cannot be used; in particular, it is not possible to associate a specific and unique identifying code with a query substructure which could then be used to identify those compounds that contained it. Instead, a number of non-unique characteristics are considered for matching purposes, these screens being structural characteristics, the presence of which is necessary, but not sufficient, for a molecule to contain the query substructure.

Substructure searching is normally effected by a two-stage search procedure. In the first stage, a screen search is carried out to identify that small sub-set of the file which might possibly match the query; this sub-set then undergoes a detailed atom-by-atom graph matching search in which each and every atom and bond in the query substructure is mapped onto the atoms and bonds of a structure in the file to determine whether a subgraph isomorphism is present. Such matching operations are very time-consuming, even if heuristics of various sorts are adopted to rapidly reject non-mappings (Sussenguth, 1965; Figueras, 1972; von Scholley, 1985), and thus most research on substructural retrieval has focussed upon the development of efficient and effective screening methodologies.

Many types of structural feature may be used to characterise chemical structures for retrieval purposes. Early screening systems adopted the manually-derived fragmentation codes described in Section 1.2.1 as the basis for automatic searching; an example of such a scheme is described by Craig and Ebert (1969). However, this approach has increasingly been replaced by fragmentation procedures which operate upon connection table records to produce screens by algorithmic means. A commonly used example of such an algorithmic fragment is the <u>augmented atom</u>, which consists of an atom together with the types of each immediately adjacent atom and bond. Thus the structure and connection table shown in Section 1.2.3 will give rise to the augmented atoms:

```
C-C     C-C-C     O-C     C-C-C     C-C=N     C=N
          |
          O
```

If an augmented atom is generated for each and every non-hydrogen atom in a compound, a relatively detailed molecular characterisation may be obtained at little computational cost. Other types of fragment descriptor may also be obtained readily from connection tables as is evidenced by the work of Lynch and his co-workers (Crowe et al., 1970; Adamson et al., 1971; Adamson et al., 1973c) who have described hierarchic series of algorithmic fragment types that are centred upon the atoms, the bonds or the rings of a structure. For example, Adamson et al. (1971) discuss a hierarchy of atom-centred fragments comprising the four following types of substructure:

```
                                              C
                                    |         |
    C            C3                -C-      C-C-N

  Simple     Coordinated         Bonded    Augmented
   atom         atom              atom       atom
```

Comparable work has been described by Feldman and Hodes (1975) who discuss the generation of fragments that are based upon chains of atoms. An example of the wide range of fragment types that may be needed for an operational system is provided by the fragment dictionary used for screening purposes in the CAS ONLINE structure and substructure search system (Dittmar et al., 1983).

Fragments such as those studied by Adamson et al. are typically quite small in extent, but highly disparate in occurrence with a few fragments occurring very frequently indeed. Thus, if simple atoms are considered, i.e. elemental occurrences, carbon atoms (72%), together with oxygen (14%) and nitrogen (7%), account for approximately 93% of the total number of non-hydrogen atoms in a typical file of organic molecules (Adamson et al., 1973b). Fluorine, sulphur, chlorine, phosphorus, bromine, silicon and iodine account for the great majority of the remaining occurrences. Such highly skewed, or <u>Zipfian</u>, frequency distributions have been identified in a wide range of information retrieval contexts (Fairthorne, 1969; Lynch, 1977).

The existence of a Zipfian distribution suggests an attractive approach to the design of screening systems if a hierarchic algorithmic fragmentation procedure is available. Very frequently occurring species are characterised by screens of high specificity while those of low occurrence are assigned more general, less discriminating screens. Thus, in the case of the atom-centred fragment hierarchy above, carbon atoms in a file will generally be characterised by augmented atom screens, whereas simple atoms would suffice to screen for the presence of a rare metal in a compound. Molecules containing atoms that occur with moderate frequency, e.g. silicon or phosphorus, could then be characterised by screens of intermediate size such as the coordinated or bonded atom levels of description.

Such a procedure implies that the selection of screens for substructure searching is based upon an analysis of the frequencies of occurrence of fragments in the file that is to be searched, and presumes that there is a relatively close match between the frequencies of substructures in the file and in the queries that are put to the search system (Adamson et al., 1973a). The resulting frequencies are then input to an algorithmic selection procedure which attempts to identify those fragments which will be most discriminating for search purposes. Factors that need to be taken into account in such a procedure include the desirability of ensuring that the resulting screens should occur approximately equifrequently in the structure file, and that the frequencies of assignment should be statistically independent of each other (Adamson et al., 1973b; Feldman and Hodes, 1975; Hodes, 1976; Willett, 1979). The fragments which are selected to act as screens are listed in a fragment coding dictionary. Then, when a new molecule is added to the file, the connection table representing the compound is analysed to identify the fragments in the coding dictionary which are present in the connection table: these fragments are then used to represent the molecule for searching purposes. A comparable procedure is used to assign screens to queries for substructure searching.

1.3.3 Graphics-based chemical retrieval software packages

Chemical search systems increasingly involve the use of sophisticated graphical input and display routines which allow the chemist to access structural data bases by means of 2-D chemical structure diagrams. The input is accomplished in such graphics-based systems by means of a light pen or stylus which the chemist uses to sketch the (sub)structure of interest on a touch-sensitive terminal or digitising tablet: software in the terminal then automatically converts the diagram into the internal connection table representation. The efficiency of this procedure may be significantly enhanced by allowing the chemist to create, amend or delete portions of the structure diagram using menu-driven software which provides pre-drawn templates for commonly-occurring structural features, such as ring skeletons or functional groups; additional menus may then be invoked to specify heteroatoms, multiple bonds, or lists of acceptable variants. Sophisticated examples of graphic structure input and display are provided by the COUSIN system developed at the Upjohn Company (Howe and Hagadone, 1982) and by the CAS ONLINE system (Dittmar et al., 1983). Once a query has been specified, the (sub)structure search routines are used to match the query against the structure file, and the retrieved molecules are then displayed on the screen of the terminal.

Several software packages for chemical searching are now available commercially. While there are similarities between them, there are also considerable differences in the means by which the basic retrieval functions are implemented, and the choice of an individual package will depend crucially upon the particular context in which it is to be used. Thus, a requirement for a fast response time may suggest the need for an inverted file organisation, which may attract significant updating and storage overheads in some cases, or for special-purpose hardware if the simpler serial file organisation is to be used; a requirement for some degree of generic searching capability will necessitate a sophisticated query language which may well prove to be too complex for many chemists to use; a requirement for a stereochemical searching capability implies more complex retrieval algorithms and the need to include stereochemical information in the connection tables which are to be searched. A discussion of these and other points is given by Ash et al. (1985) who present a detailed comparative review of six different chemical structure searching systems.

While early chemical information systems were set up primarily to support structure and substructure searching, the ease with which machine-readable files of structural data can be manipulated soon led to an interest in the use of structural information for other purposes. In particular, extensive use is now made of techniques for the correlations of molecular structure with biological, physical and chemical properties of various sorts. This work is reviewed in the following section.

1.4 QUANTITATIVE STRUCTURE-ACTIVITY RELATIONSHIP STUDIES

1.4.1 Introduction

A range of methods is currently used for the discovery of novel compounds with desirable characteristics, e.g., pesticidal or anti-hypertensive activity. The methods include the systematic modification of naturally occurring compounds, the investigation of unexpected side-effects or metabolites of known drugs, studies of folk medicine, and even serendipity (Franke, 1984). Even so, many thousands of molecules may need to be synthesised and tested before a viable compound can be brought to the market place, and there has accordingly been considerable interest in the use of computational methods which can increase the efficiency of discovery of molecules with required biological activity.

An obvious means of providing computational support is the use of quantum mechanical techniques which permit the calculation of a wide range of important molecular properties, such as dipole moments, equilibrium geometries, ionisation potentials and charge distributions. Unlike the techniques discussed below, quantum mechanics has a rigorous theoretical basis and can be used for the description of a range of phenomena in much greater detail than can be achieved by alternative approaches (Christoffersen, 1979; Boyd, 1983). However, the methods are generally extremely demanding of computational resources, and often require extensive parameterisation to achieve even moderately accurate results. Accordingly, greater reliance is placed on quantitative structure-activity relationship (QSAR) techniques which involve the correlation of physical, biological and chemical property data with variables characterising the structures in a data set. The establishment of a statistically significant correlation equation provides a rational approach to drug design since it serves to focus attention on those classes of molecule that have high a priori probabilities of activity. QSAR techniques are semi-empirical in nature, and it is these simpler methods that are described in the remainder of this section. General reviews of the field are given by Martin (1981), Topliss (1983) and Franke (1984).

QSAR methods can be used to aid drug discovery programmes in two ways. The first, and historically the most important, has been the use of lead optimisation techniques. Here, a previously discovered lead compound, i.e., one which has been shown to exhibit the property of interest, is subjected to carefully controlled chemical variation, the aim being to arrive at a member of the congeneric series which exhibits the best trade-off between the required activity, undesirable side-effects, and other factors such as ease of preparation and solubility characteristics. Examples of QSAR techniques that may be used for lead optimisation include Hansch analysis, Free-Wilson analysis and molecular graphics.

The use of QSAR for lead optimisation is now widely practised. Much less common is the use of QSAR for lead generation, i.e. the identification of a new lead compound that can then form the basis for a subsequent optimisation programme. This area is, at present, more problematic,and is currently much less used for the identification of novel leads than the traditional methods mentioned above. Substructural analysis methods are the most appropriate for lead generation, and provide a means of suggesting compounds with potential activity and of identifying substructural features that are important for activity.

Lead generation methods are normally used with the activity data resulting from empirical or random screening, in which a simple but effective biological test is used to eliminate from further consideration great numbers of compounds that cannot possibly possess the desired activity characteristics. Such screens are thus comparable in nature to the chemical screens which are used for substructure searching as described in Section 1.3. Biological screens are typically capable of processing a high throughput of new molecules, and are often arranged to allow through a pre-defined fraction of the tested compounds for further, and more detailed, testing.

Techniques for QSAR may be classified on the basis of the number of compounds that can be processed in a single analysis. Thus, the computationally demanding quantum mechanics and molecular graphics approaches are generally restricted to small classes of compounds which can be analysed in very great detail, while the common Free-Wilson, Hansch and pattern recognition methods are typically applied to data sets containing a few tens or hundreds of compounds. Substructural analysis techniques may be used for the processing of large data sets, containing many thousands of structures in some cases. An effective drug development programme will involve several different QSAR methods, and will operate most efficiently when the simpler techniques are applied to the largest volumes of data: this allows the identification of a small number of compounds that can then be passed on for processing by more sophisticated techniques.

Studies of QSAR involve two major factors: the selection of the variables that are used to encode the structural characteristics of the compounds in the data set, and a mathematical expression which relates these variables to the observed property data which can be of two types. Quantitative data requires the measurement of some numerical property on an interval or ratio scale. A common approach is to measure the concentration, C, of compound that is required to produce a specified biological response, e.g., LD_{50}s which are used in toxicological testing; activity is then expressed as log(1/C). Qualitative data is measured on an ordinal scale, and thus a compound might be coded as highly active, moderately active or inactive. In many cases, only binary data is used, i.e. active or inactive. Such data is characteristic of most biological screening systems. The type of data available may influence the choice of mathematical function which is used to formulate the QSAR. For example, multiple regression techniques are appropriate only to the analysis of quantitative data, and thus QSAR methods based on this technique may not be appropriate for the analysis of much screening data, which may be processed using non-parametric techniques such as linear learning machines.

1.4.2 Hansch analysis

By far the most popular QSAR technique is the physicochemical or extrathermodynamic method in which physicochemical parameters are correlated with the observed biological activities of groups of congeneric structures using multiple regression. Due to the detailed studies that have been made by Hansch and his co-workers over the years, the technique is often referred to as Hansch analysis (Hansch, 1969; Hansch, 1981; Franke, 1984).

The use of physicochemical parameters for drug design stems from the pioneering studies of Hammett and Taft

(Hansch, 1981). Hammett investigated the electronic effects of aromatic ring substituents, and introduced the idea of a numerical scale that could be used to denote the contribution to reactivity of a given substituent. The resulting substituent constants have been found to be of great help in rationalising electronic effects in many different types of organic reactions, and the success of the approach led Taft to develop a comparable steric parameter. Hansch extended the scope of the physicochemical approach to QSAR studies by emphasising the importance of lipophilic properties in determining the level of biological activity. The parameter that has been chosen to measure the overall lipophilicity of a molecule is the logarithm of the octanol/water partition coefficient, logP. In many investigations, the hydrophobic nature of individual parts of molecules is of interest, rather than the nature of the entire structure, and in such cases, individual substituent values may be used to obtain a good correlation. Such values are found, in large part, to be additive in character, and this has formed the basis for computer programs that permit the calculation of logP values for molecules where the partition coefficient has not been determined experimentally (Chou and Jurs, 1979). A linear relationship between activity and logP is commonly observed, but quadratic relationships may also be obtained in many cases (Hansch and Clayton, 1973). The fourth major parameter that has been found to be of wide applicability in QSAR studies is the molar refractivity which represents the volume of a compound corrected by the ease of polarisability of its electrons; as with logP, individual substituent values may be used if required.

Physicochemical QSARs are obtained by means of regression analysis packages which fit the observed activity data with the parameter values for those compounds for which property data is available, the training set. The regression equation may then be used to predict the activity of untested structures in the prediction set. One or more of the chosen physicochemical parameters may be needed to obtain a good correlation for a set of structures, as well as indicator variables, binary variables which indicate the presence or absence of certain specific molecular features; the use of such variables emphasises the close relationship between Hansch analysis and the substructural analysis techniques described in the next section.

Hansch analysis is restricted in that it should be used only when all of the molecules in a data set exhibit activity via the same mode of action, and Hansch analysis is thus used as an optimisation technique to maximise activity within a congeneric series. Other limitations include problems of statistical validity (Topliss and Edwards, 1979) and the need to obtain a reasonable spread of parameter values (Wooton et al., 1975). However, a very large number of analyses has now been carried out using the approach, and there is thus a considerable body of evidence on how the various sorts of physicochemical parameter may best be employed in QSAR investigations. Indeed, the extensive use of the technique has meant that it is now almost synonymous with QSAR.

1.4.3 Substructural analysis

The ease with which structural data can be manipulated in chemical information systems has led to a rapid increase in the use of QSAR methods where substructural occurrence data is correlated with activity. Such an approach derives in large part from the pioneering study by Free and Wilson (1964), and takes as its basic postulate the assumption that a given substructure makes an additive and constant contribution to the overall activity of a molecule. Free and Wilson suggested the approach for lead optimisation within congeneric series, and adopted the additional constraint that a particular substructural contribution to activity was dependent upon its position of substitution on the basic ring structure of the series. Then, regression analysis may be used to determine the effect of each possible substituent at each possible substitution position. Thus, if P[I,J] is used to indicate the presence (P[I,J]=1) or absence (P[I,J]=0) of the I'th substructure at the J'th substitution position, the regression analysis permits the calculation of A[I,J], the contribution to activity of the I'th substructure at the J'th position. The activity of a molecule in the prediction set may then be estimated by setting the various P[I,J] values to 0 or 1, depending upon which substructures are present and at which locations. The approach provides a simple and direct formulation of the relationship between structure and activity but suffers from the limitations that it can only be used where multiple substitution sites exist, and where there is a minimum of two occurrences for each combination of substructure and position: this severely restricts the practical application of the method, although modifications have been suggested which can overcome these problems (Fujita and Ban, 1971).

The use of regression analysis in both Hansch and Free-Wilson analysis to identify contributions to the overall activity suggests a similarity of approach, and it is indeed possible to describe Free-Wilson activity contributions and Hansch substituent constants in terms of each other (Franke, 1984). Both types of analysis are constrained by the fact that they are intended primarily for use with structurally related groups of molecules.

Structurally diverse data sets may be processed using the methods of substructural analysis, which is related to Free-Wilson analysis and to the additive modelling techniques which have long been used for the estimation of thermochemical and physical properties. Substructural analysis removes the positional character of the activity contributions in Free-Wilson analysis, i.e., it assumes the contribution to activity of a given substructure is constant, irrespective of the type of compound in which it occurs. Whilst crude in the extreme, such an approach has been shown to allow the identification of structure-activity relationships with data sets which may be large, or very large, in size, widely disparate in nature, and characterised by both qualitative and quantitative activity data. This work is summarised in the remainder of this section.

The studies of Adamson and his co-workers are perhaps the most closely related to the traditional Free-Wilson approach. This group has investigated a wide range of data sets for which quantitative activity data is available, and for which the molecules are characterised by fragment substructures generated automatically from either connection table (Adamson and Bush, 1974) or WLN (Adamson and Bawden, 1975) representations of molecular structure. In the former case, the descriptors are the small atom-centred and bond-centred fragment screens described in Section 1.3, while in the latter they are WLN symbol strings representing various types of chemical functionality. Regression analysis is again used for the calculation of the individual fragment contributions, and hence for the prediction of activity in untested compounds. The use of WLN provides a simple and computationally efficient means of taking the relative positions of different fragments into account when setting up the regression equation (Adamson et al., 1984), and thus provides a mid-point between the detailed positional information that is required for Free-Wilson analysis and the complete neglect of such information that characterises most substructural analysis techniques.

Cramer et al. (1973) have described a substructural approach that may be used for the analysis of qualitative property data. If the I'th substructure occurs in N[I] compounds, NAC[I] of which are active, then the ratio NAC[I]/N[I] provides a measure of the probability that a compound containing this feature will be active. The overall activity of a compound is obtained by calculating the average probability for all of the substructures that are present in it, and these mean activities thus provide a mechanism for the <u>ranking</u> of a set of compounds. Molecules that occur near the top of such a ranking are those that contain many fragments with high activity ratios, and are thus those with the highest probability of activity in the biological test of interest. Hodes and his co-workers (Hodes et al., 1977; Hodes, 1981) have extended this basic model by using a statistically defined activity weighting scheme which allows the calculation of both activity and inactivity weights for the fragments substructures; a system based on such ideas is now in regular use to assist in the selection of compounds for carcinogenicity testing at the National Cancer Institute. A further modification has been described by Carhart et al. (1985) which can be used with both qualitative and quantitative activity data.

Rather than using multiple regression or ranking methods to obtain predictions of activity, <u>pattern recognition</u> methods attempt to identify a <u>classification</u> function which can assign molecules to one of two, or more, activity classes (Stuper et al., 1979). Compounds are represented as points in a multi-dimensional space, where each dimension corresponds to one of the set of descriptors that is used to characterise the structures in the data set. The expectation is that compounds with comparable activities will cluster together in the same region of the space while being distanced from compounds with different activities: normally, only two classes of activity are considered, i.e. active or inactive. Two main types of classification function can be used for the classification of compounds in the prediction set. The first of these involves the identification of a hyperplane which divides the space so that all of the active molecules lie on one side of the plane and all of the inactives on the other. Alternatively, a distance classifier involves determining the distance, or similarity, of the unknown from the actives and inactives in the training set. Classification with a hyperplane is obtained by plotting the unknown in the hyperspace and then determining upon which side of the hyperplane it lies: if it is found to lie upon the side of the hyperplane which corresponds to the actives in the training set it is predicted to be active. The most common way of obtaining such hyperplane separators is the linear learning machine. In this, an initial plane is iteratively modified by the classification of compounds in the training set until all, or most, of them can be classified correctly. The resulting surface can then be used for the classification of molecules in the prediction set. Such an approach is intuitively appealing and has a well-established body of mathematical theory to support it, but there are at least two limitations. Firstly, the learning procedure may need to be iterated many hundreds or thousands of times before it converges to the final hyperplane; secondly, the hyperplane which is identified is not necessarily unique since its derivation is dependent upon the order in which the members of the training set have been processed.

The most common distance classifier is the K-nearest neighbour method which involves determining the distance of the unknown from each of the compounds in the training set so as to identify the K molecules which are closest to it. K is normally an odd number, so that the predicted activity for the unknown may be obtained by a majority vote of the neighbours; alternatively, decision functions may be used which consider the actual magnitudes of the various distances (Franke, 1984). Distance classifiers are simple in concept and can be used with data sets for which it is not possible to identify a hyperplane separator; however, they do require the measurement of the distance to each and every member of the training set unless some nearest neighbour searching strategy is used to reduce the number of distance calculations. Efficient algorithms for nearest neighbour searching are described in Chapter 5.

Pattern recognition methods were first used in chemistry for the assignment of compounds to structural classes using mass spectral data (Jurs and Isenhour, 1975) but have more recently been applied to QSAR problems (Stuper et al., 1979). There have been several criticisms of such studies on grounds of statistical validity, the criteria that are used for the selection of the substructural features, and the distribution of active and inactive compounds in the data sets studied (Wold and Dunn, 1983). However, an awareness of the potential problems should ensure the appropriate use of such techniques in the future.

An obvious question that must be addressed in any

substructural analysis investigation is what types of structural feature are to be considered. The availability of facilities for structure and substructure searching has often led to fragment screens being used as molecular descriptors; as an example, augmented atom fragments have been used by several workers, and also form the primary type of molecular descriptor considered in the work reported in this book. However, it is by no means clear that fragments which were originally derived for information retrieval purposes should be equally applicable to QSAR studies. Accordingly, a wide range of other types of structural feature has been used, these including ring descriptors of various sorts automatically extracted from WLNs or from connection tables, manual fragmentation codes, algorithmic definitions of chemical functionality, connected sequences of atoms or bonds, and connectivity indices which describe the degree of branching within a structure. In many cases, a range of such descriptor types may be used within a single investigation, with the structural features possibly augmented by physicochemical parameters and geometric features; an example of such a system is described by Brugger et al. (1979), while detailed reviews of structural features for QSAR are presented by Bawden (1983) and by Franke (1984).

Of the various types of QSAR method discussed in this section, substructural analysis methods are the most closely related to the fragment-based similarity and clustering techniques presented in subsequent chapters. However, there are differences between the two types of technique. Firstly, while similarity measures can indeed be used to rank compounds, as described in Chapter 3, the ranking does <u>not</u> require the prior analysis of fragment occurrences in active and inactive compounds which is needed in substructural analysis; instead, the ranking of a whole series of unknown compounds may be obtained by reference to a single compound of known activity. This is the exact inverse of a K-nearest neighbour classification, where a single unknown is compared with all of the members of the training set. Secondly, while pattern recognition involves the classification of compounds, this procedure involves just the assignment of molecules to pre-defined classes; cluster analysis, or automatic classification, requires the identification of the classes that are present within a data set before such assignments can be carried out.

1.4.4 Molecular graphics systems

The QSAR methods described in the previous sections are constrained by the fact that they generally take little or no account of stereochemical and spatial information, and it may often be difficult to relate actions at the molecular level with the observed mathematical formulations of the structure-activity relationship. For this reason, attention is now focussing upon the use of molecular graphics techniques which provide an insight into the 3-D interactions between potential drug molecules and biological receptor sites (Feldman, 1976; Langridge et al., 1981). The basic rationale for the consideration of 3-D information is the principle that a drug 'fits' its biological receptor much as a key fits a lock, although the idea of a perfectly rigid drug fitting into a perfectly rigid receptor is only a convenient approximation to reality.

A primary requirement for a molecular graphics system is the provision of structural information in a convenient and accessible form, and the most obvious source of such data is from the crystallographic structure determinations itemised in the Cambridge Crystallographic Data Bank (Allen et al., 1979). In many cases, of course, the molecule of interest has not been analysed crystallographically, and in this case a molecular building routine will be required to construct a whole molecule from small fragments automatically, as is done manually with Dreiding or CPK models; alternatively, a 2-D structure diagram may be input using a light pen or structure display program as with the systems described in Section 1.3.3 (Gund et al., 1980). The resulting connection table is then used as the input to a molecular mechanics procedure, which seeks to minimise the energy within the molecule (Boyd and Lipkowitz, 1982), or to the distance geometry approach of Wenger and Smith (1982). Alternatively, stable conformations may be identified using quantum mechanical methods; however, these are very much more time consuming than molecular mechanics routines which can typically identify the stable conformations of a low molecular weight compound in a few minutes of CPU time using modern minicomputer systems.

The primary function of a molecular graphics systems is the provision of easy-to-use and powerful 3-D display facilities so as to allow the medicinal chemist an insight into the spatial disposition of the atoms and bonds within a structure. A range of monochrome and coloured display routines are normally available, including ball-and-stick, space-filling formats, and stereoscopic pair line diagrams. With increasing hardware power and software sophistication, real-time molecular rotation of such displays may be obtained which provides a striking means of revealing the full molecular structure. In addition to the manipulation of individual molecules, routines are often provided to allow for the comparison and superimposition of pairs, or larger numbers, of structures to identify common features and for the approach of a molecule to a receptor site, the motion being controlled by joysticks or thumb-wheels which are manipulated by the medicinal chemist who is running the program (Langridge et al., 1981).

A further major application of molecular graphics systems is in the analysis and display of pharmacophores, the set of structural features in a drug molecule that is recognised at the receptor site. Pharmacophores may be defined in both topological or topographical terms, but the latter provide so much more information about the exact molecular orientation required for activity that most attention has been devoted to such 3-D patterns. The most widely used method of pharmacophoric pattern identification is to compare the energetically feasible conformations of those molecules which are known to exhibit the desired activity so as to obtain a map of the receptor site. Molecules that satisfy the spatial requirements of this site may then be expected to exhibit the desired activity: detailed accounts of such methods are given by Gund (1977) and by Humblet and Marshall (1981). Once the pharmacophore has been identified, automatic means are needed to determine whether it is present in a molecule and there have been reports in the literature of algorithms for the 3-D analogue of 2-D substructure searching (Gund, 1977; Jakes and Willett, 1986).

1.5 OTHER APPLICATION AREAS

The two previous sections have attempted to give an overview of the current state of development of techniques for structure retrieval and QSAR. As noted in Section 1.1, it is in these two areas that the techniques presented later in the book are likely to have their most obvious application. However, the chemical structure diagram can act as a link between many different types of information system, and this section briefly describes other facilities which may be found in modern chemical information systems.

Computerised systems for the storage and retrieval of chemical reaction information provide an obvious and natural complement to the structure-based systems described in Section 1.3. However, the development of reaction information systems has been very slow, principally because of the lack of an efficient and effective reaction indexing procedure (Willett, 1986). Whereas a molecule may be characterised by the fragments contained within it, the focus of interest in a reaction is the parts of the reactant and product molecules which have been involved in the change, i.e. the substructural transformation which has taken place. The automatic identification of such <u>reaction sites</u> involves a comparison of the connection tables of the reacting molecules to identify features that are common to the two sides of the reaction equation and hence, by inference, the areas of difference. As with structure and substructure searching, a graph-theoretic isomorphism procedure is available for the identification of reaction sites: however, such <u>maximal common subgraph</u> algorithms are even more demanding of computational resources than are graph and subgraph isomorphism algorithms (McGregor and Willett, 1981), and it is only within the last few years that software packages have become available for the handling of reaction information (Willett, 1986).

A more sophisticated use of reaction information

underlies computer-aided synthesis design (CASD) programs. Such programs assist the chemist in his search for viable syntheses by suggesting potential routes to the target compound, i.e. the molecule whose synthesis is required. CASD programs can be either empirical or non-empirical in nature (Bersohn and Esack, 1976; Wipke and Howe, 1977).

The more common empirical programs make use of libraries of known reactions which are accessed to determine whether any of them might be used to assist in the synthesis of the target structure. Interactive programs accept as input the molecule which is to be synthesised, and output precursor molecules which could be transformed into the target by a single reaction, although such a reaction might involve several separate synthetic steps. The chemist running the program can then inspect the precursors and select one or more of them as the basis for the next iteration, thus creating a chain of reactions in the retro-synthetic direction: the procedure terminates when a readily available precursor is identified. Alternatively, non-interactive programs proceed without intervention by the chemist, with the selection of precursors being carried out automatically.

Non-empirical programs do not involve a reaction library, but are instead based upon the abstract manipulation of sets of bonds and electrons. Such programs need not be constrained by current chemical knowledge and are thus able to predict synthetic routes in previously unexplored areas of chemistry.

CASD and reaction retrieval systems are complementary in nature; the former involve sophisticated and computationally demanding operations on libraries of, typically, a few hundreds of reactions (in the case of the empirical programs), while the latter involve relatively simple characterisations of data bases that may contain many tens of thousands of reactions. Accordingly, there is now increasing interest in the integration of these two types of reaction information processing system (Ash et al., 1985).

Computer-aided structure elucidation (CASE) involves the machine identification of an unknown molecule on the basis of analytical evidence, typically spectroscopic data of various sorts (Smith, 1977; Lindsay et al., 1980). A CASE program will typically have a data base of the spectral patterns associated with each of a library of standard substructures, and thus the identification of a particular spectral line, or series of lines, may be used to infer the presence of the corresponding substructure in the molecule under investigation. Once all, or most, such substructures have been identified, a structure generation routine is used to generate all chemically feasible molecules that satisfy the substructural constraints that have been identified during the analysis of the spectrum. After validation, the spectrum is predicted for each feasible molecule and compared with the experimental spectrum to determine which molecules have the highest probability of being the unknown. This cycle of analysis, generation and comparison may need to be iterated several times until an acceptable number of possible compounds has been identified.

REFERENCES

Adamson, G.W. and Bawden, D. (1975). A method of structure-activity correlation using Wiswesser Line Notation. J. Chem. Inform. Comput. Sci. 15, 215-220.
Adamson, G.W., Bawden, D. and Saggers, D.T. (1984).

Quantitative structure-activity relationship studies of acute toxicity (LD_{50}) in a large series of herbicide benzimidazoles. Pestic. Sci. 15, 31-39.

Adamson, G.W. and Bush, J.A. (1974). A method for relating the structure and properties of chemical compounds. Nature 248, 406-407.

Adamson, G.W., Clinch, V.A. and Lynch, M.F. (1973a). Relationship between query and data-base microstructure in general substructure search systems. J. Chem. Docum. 13, 133-136.

Adamson, G.W., Cowell, J., Lynch, M.F., McLure, A.H.W., Town, W.G. and Yapp, A.M. (1973b). Strategic considerations in the design of a screening system for substructure searches of chemical structure files. J. Chem. Doc. 13, 153-157.

Adamson, G.W., Creasey, S.E., Eakins, J.P. and Lynch, M.F. (1973c). Analysis of structural characteristics of chemical compounds in a large computer-based file. Part 5. More detailed cyclic fragments. J. Chem. Soc. (C), 2071-2076.

Adamson, G.W., Lynch, M.F. and Town, W.G. (1971). Analysis of structural characteristics of chemical compounds in a large computer-based file. Part 2. Atom-centred fragments. J. Chem. Soc. (C), 3702-3706.

Allen, F.H., Bellard, S., Brice, M.D., Cartwright, B.A., Doubleday, A., Higgs, H., Hummelick, T., Hummelick-Peters, B.G., Kennard, O., Motherwell, W.D.S., Rodgers, J.R. and Watson, D.G. (1979). The Cambridge Crystallographic Data Centre: computer-based search, retrieval, analysis and display of information. Acta Crystallograph. B35, 2331-2339.

Ash, J.E., Chubb, P.A., Ward, S.E., Welford, S.M. and Willett, P. (1985). Communication, Storage and Retrieval of Chemical Information. Ellis Horwood, Chichester.

Ash, J.E. and Hyde, E. (eds.) (1975). Chemical Information

Systems. Ellis Horwood, Chichester.
Barnard, J.M. (ed.) (1984). Computer Handling of Generic Chemical Structures. Gower, Aldershot.
Bawden, D. (1983). Computerized chemical structure-handling techniques in structure-activity studies and molecular property prediction. J. Chem. Inform. Comput. Sci. 23, 14-22.
Bawden, D., Catlow, J.T., Devon, T.K., Dalton, J.M., Lynch, M.F. and Willett, P. (1981). Evaluation and implementation of topological codes for online compound search and registration. J. Chem. Inform. Comput. Sci. 21, 83-86.
Bersohn, M. and Esack, A. (1976). Computers and organic synthesis. Chem. Rev. 76, 269-282.
Boyd, D.B. (1983). Quantum mechanics in drug design: methods and applications. Drug Inform. J. 17, 121-131.
Boyd, D.B. and Lipkowitz, K.B. (1982). Molecular mechanics. The method and its underlying philosophy. J. Chem. Educ. 59, 269-274.
Brugger, W.E., Stuper, A.J. and Jurs, P.C. (1976). Generation of descriptors from molecular structure. J. Chem. Inform. Comput. Sci. 16, 105-110.
Carhart, R.E., Smith, D.H. and Venkataraghavan, R. (1985). Atom pairs as molecular features in structure-activity studies: definition and applications. J. Chem. Inform. Comput. Sci. 25, 64-73.
Chou, J.T. and Jurs, P.C. (1979). Computer-assisted computation of partition coefficients for molecular structures using fragment constants. J. Chem. Inform. Comput. Sci. 19, 172-178.
Christoffersen, R.E. (1979). Quantum pharmacology: recent progress and current status. Amer. Chem. Soc. Sympos. Ser. 112, 3-19.
Craig, P.N. and Ebert, H.M. (1969). Eleven years of structure searching using the SKF (Smith, Kline and

French) fragment codes. J. Chem. Docum. 9, 141-146.

Cramer, R.D., Redl, G. and Berkoff, C.E. (1973). Substructural analysis. A novel approach to the problem of drug design. J. Med. Chem. 17, 533-535.

Crowe, J.E., Lynch, M.F. and Town, W.G. (1970). Analysis of structural characteristics of chemical compounds in a large computer-based file. Part 1. Non-cyclic fragments. J. Chem. Soc. (C), 990-996.

Dittmar, P.G., Farmer, N.A., Fisanick, W., Haines, R.C. and Mockus, J. (1983). The CAS ONLINE search system. I. General system design and selection, generation and use of search screens. J. Chem. Inform. Comput. Sci. 23, 93-102.

Fairthorne, R.A. (1969). Empirical hyperbolic distributions (Bradford-Zipf-Mandelbrot) for bibliometric description and prediction. J. Docum. 25, 319-343.

Feldman, A. and Hodes, L. (1975). An efficient design for chemical structure searching. I. The Screens. J. Chem. Inform. Comput. Sci. 15, 147-152.

Feldmann, R.J. (1976). The design of computing systems for molecular modeling. Ann. Rev. Biophys. Bioeng. 5, 477-510.

Figueras, J. (1972). Substructure search by set reduction. J. Chem. Docum. 12, 237-244.

Franke, R. (1984). Theoretical Drug Design Methods. Elsevier, Amsterdam.

Free, S.M. and Wilson, J.W. (1964). A mathematical contribution to structure-activity studies. J. Med. Chem. 7, 395-399.

Freeland, R.G., Funk, S.A., O'Korn, L.J. and Wilson, G.A. (1979). The CAS Chemical Registry System. II. Augmented connectivity molecular formula. J. Chem. Inform. Comput. Sci. 19, 94-97.

Fujita, T. and Ban, T. (1971). Structure-activity study of

phenethylamines as substrates of biosynthetic enzymes of sympathetic transmitters. J. Med. Chem. 14, 148-152.

Garfield, E., Revesz, G.S., Granito, C.E., Dorr, H.A., Calderon, M.M. and Warner, A. (1970). Index Chemicus Registry System: pragmatic approach to substructure chemical retrieval. J. Chem. Docum. 10, 54-58.

Granito, C.E., Roberts, S. and Gibson, G.W. (1972). The conversion of Wiswesser Line Notations to Ring Codes. I. The conversion of ring systems. J. Chem. Docum. 12, 190-196.

Granito, C.E. and Rosenberg, M.D. (1971). Chemical Substructure Index (CSI). A new research tool. J. Chem. Docum. 11, 251-256.

Gund, P. (1977). Three-dimensional pharmacophoric pattern searching. Prog. Molec. Sub-Cell. Biol. 5, 117-143.

Gund, P., Andose, J.D., Rhodes, J.B. and Smith, G.M. (1980). Three-dimensional molecular modeling and drug design. Science 208, 1425-1431.

Hansch, C. (1969). A quantitative approach to biological structure-activity relationships. Acc. Chem. Res. 2, 232-239.

Hansch, C. (1981). The physicochemical approach to drug design and discovery (QSAR). Drug. Develop. Res. 1, 267-309.

Hansch, C. and Clayton, J.M. (1973). Lipophilic character and biological activity of drugs. II. The parabolic case. J. Pharm. Sci. 62, 1-21.

Hodes, L. (1976). Selection of descriptors according to discrimination and redundancy. Application to chemical structure searching. J. Chem. Inform. Comput. Sci. 16, 88-93.

Hodes, L. (1981). Computer-aided selection of compounds for antitumour screening: validation of a statistical-heuristic method. J. Chem. Inform. Comput. Sci. 21, 128-132.

Hodes, L., Hazard, G.F., Geran, R.I. and Richman, S. (1977). A statistical heuristic method for automated selection of drugs for screening. J. Med. Chem. 20, 469-475.

Howe, W.J. and Hagadone, T.R. (1982). Molecular substructure searching: computer graphics and query entry methodology. J. Chem. Inform. Comput. Sci. 22, 8-15.

Humblet, C. and Marshall, G.R. (1981) Three-dimensional computer modeling as an aid to drug design. Drug Develop. Res. 1 409-434.

Jakes, S.E. and Willett, P. (1986) Pharmacophoric pattern matching in files of three-dimensional chemical structures. Selection of inter-atomic distance screens. J. Mol. Graphics 4, 12-20.

Jurs, P.C. and Isenhour, T.L. (1975). Chemical Applications of Pattern Recognition. Wiley, New York.

Langridge, R., Ferrin, T.E., Kuntz, I.D. and Connolly, M.L. (1981). Three-dimensional colour graphics in studies of molecular interactions. Science 211, 661-666.

Lindsay, R.K., Buchanan, B.G., Feigenbaum, E.A. and Lederberg, J. (1980). Applications of Artificial Intelligence for Organic Chemistry. McGraw-Hill, New York.

Lynch, M.F. (1977). Variety generation - a re-Interpretation of Shannon's's mathematical theory of communication and its implications for information science. J. Amer. Soc. Inform. Sci. 28, 19-25.

McGregor, J.J. and Willett, P. (1981). Use of a maximal common subgraph algorithm in the identification of the ostensible bond changes occurring in chemical reactions. J. Chem. Inform. Comput. Sci. 21, 137-140.

Martin, Y.C. (1981). A practitioner's perspective of the role of quantitative structure-activity analysis in medicinal chemistry. J. Med. Chem. 24, 229-237.

Maurer, W.D. and Lewis, T.G. (1975). Hash table methods. Comput. Surv. 7, 5-19.

Morgan, H.L. (1965). The generation of a unique machine-description for chemical structures - a technique developed at Chemical Abstracts Service. J. Chem. Docum. 5, 107-113.

Petrarca, A.E., Lynch, M.F. and Rush, J.E. (1967). A method for generating unique computer structural representations of stereoisomers. J. Chem. Docum. 7, 154-165.

Read, R.C. and Corneil, D.G. (1977). The graph isomorphism disease. J. Graph. Theory 1, 339-363.

Rouvray, D.H. (1977). The changing role of the symbol in the evolution of chemical notation. Endeavour 1, 23-31.

Rush, J.E. (1976). Status of notation and topological systems and potential future trends. J. Chem. Inform. Comput. Sci. 16, 202-210.

Smith, D.H. (1977). Computer-Assisted Structure Elucidation. American Chemical Society, Washington.

Smith, E.G. and Baker, P.A. (1975). The Wiswesser Line-Formula Chemical Notation (WLN) Chemical Information Management Inc., New Jersey.

Stuper, A.J., Brugger, W.E. and Jurs, P.C. (1979). Computer-Assisted Studies of Chemical Structure and Biological Function. Wiley, New York.

Sussenguth, E.H. (1965). A graph-theoretic algorithm for matching chemical structures. J. Chem. Docum. 5, 36-43.

Tarjan, R.E. (1977). Graph algorithms for chemical computation. Am. Chem. Soc. Sympos. Ser. 36, 1-19.

Topliss, J.G. (1983). Quantitative Structure-Activity Relationships of Drugs. Academic Press, New York.

Topliss, J.G. and Edwards, R.P. (1979). Chance factors in QSAR studies. Amer. Chem. Soc. Sympos. Ser. 112, 131-145.

Vander Stouw, G.G., Elliott, P.M. and Isenberg, A.C.

(1974). Automated conversion of chemical substance names to atom-bond connection tables. J. Chem. Docum. 14, 185-193.

Vander Stouw, G.G., Gustafson, C., Rule, J.D. and Watson, C.E. (1976). The Chemical Abstracts Service Chemical Registry System. IV. Use of the Registry System to support the preparation of index nomenclature. J. Chem. Inform. Comput. Sci. 16, 213-222.

von Scholley, A. (1985). A relaxation algorithm for generic chemical structure screening. J. Chem. Inform. Comput. Sci. 25, 235-241.

Wenger, J.C. and Smith, D.H. (1982). Deriving three-dimensional representations of molecular structure from connection tables augmented with configuration designations using distance geometry. J. Chem. Inform. Comput. Sci. 22, 29-34.

Willett, P. (1979). A screen set generation algorithm. J. Chem. Inform. Comput. Sci. 19, 159-162.

Willett, P. (1980). Nomenclature processing and the interconversion of chemical structure representations. Proceedings of the CNA(UK) Seminar on Chemical Structure Searching of the Published Literature. Chemical Notation Association (UK), London.

Willett, P. (1986). Modern Approaches to Chemical Reaction Searching. Gower, Aldershot.

Wipke, W.T. and Dyott, T.M. (1974). Stereochemically unique naming algorithm. J. Amer. Chem. Soc. 96, 4825-4834.

Wipke, W.T. and Howe, W.J. (1977). Computer-Assisted Organic Synthesis. American Chemical Society, Washington.

Wold, S. and Dunn, W.J. (1983). Multivariate quantitative structure-activity relationships (QSAR): conditions for their applicability. J. Chem. Inform. Comput. Sci. 23, 6-13.

Wooton, R., Cranfield, R., Sheppey, G.L. and Goodford, P.J. (1975). Physicochemical-activity relationships in practice. 2. Rational selection of benzenoid substituents. J. Med. Chem. 18, 607-613.

CHAPTER 2
Similarity and Cluster Analysis

2.1 INTRODUCTION

Clustering involves the grouping of similar objects, and has been practised, consciously or unconsciously, for many thousands of years. The development of computer technology has resulted in cluster analysis, the name which is given to techniques that allow the production of a range of different types of classification by purely automatic means. Other names include numerical taxonomy, systematics and botryology (Hartigan, 1975). Cluster analysis is often referred to as classification but this is potentially misleading. Classification normally refers to the assignment of objects to pre-defined classes, whereas clustering allows the identification of these classes: thus clustering must precede classification in the analysis of a data set. Cluster analysis was first studied intensively in the biological sciences, but is now used in a wide range of disciplines including archaeology, astronomy, computer science, geology, information science, market research, geology, marketing, medicine and psychiatry (Cormack, 1971; Sneath and Sokal, 1973; Everitt, 1974; Hartigan, 1975; Dubes and Jain, 1980; Spath, 1980; Lee, 1981; Gordon, 1981). In this book, we discuss the use of cluster analysis techniques to increase further the range of facilities available in computerised chemical information systems. Specifically, in Chapters 3 and 4 we describe novel methods for the searching and grouping of machine-readable files of chemical structures characterised by fragment substructures, while Chapter 5 presents algorithms for the efficient implementation of these techniques. This chapter gives an introduction to cluster analysis, with special attention to those parts of the subject that are considered in more detail in the experimental studies presented later in the book. The chapter commences by discussing the calculation of similarity between pairs of objects, an important component of any clustering procedure. Cluster analysis can result in the production of either hierarchic or non-hierarchic classifications, and clustering methods for these two types of classification are described in Sections 2.3 and 2.4. Section 2.5 summarises work to date on the comparison and evaluation of clustering methods, a topic that is considered in detail in the chemical context in Chapter 4, and the chapter concludes with an introduction to some of the ways in which clustering techniques can be used in chemical information systems.

2.2 THE CALCULATION OF SIMILARITY

The methods of cluster analysis which are described later in this chapter are all based in some way upon measurements of the similarity between a pair of objects. In what follows, it will generally be assumed that the similarity is being measured between one object and another; however, analogous arguments may be used to describe the similarity between an object and a cluster, as occurs in the relocation clustering methods of Section 2.4.2, or between two clusters, as occurs in the hierarchic agglomerative clustering methods of Section

2.3.2.

This section describes the calculation of similarity with special reference to techniques that are appropriate to chemical structure handling system. In general, the determination of similarity between a pair of objects involves three major steps, these being the selection of the variables that are to be used to characterise the objects, the selection of a weighting scheme for these variables, and the selection of a similarity coefficient to determine the degree of resemblance between the two attribute vectors.

In the chosen application area of chemical information systems, the first of these steps, the identification of possible descriptor types, is not a problem since it is assumed that molecular characterisations are available in the shape of the simple, fragment-based representations described in Chapter 1. Accordingly, the discussion that follows concentrates upon weighting schemes and similarity coefficients which may be used for the determination of inter-molecular similarity. An extremely detailed account of the measurement of similarity is given by Sneath and Sokal (1973), and the reader is referred to this source for further details of the similarity measures discussed here.

2.2.1 Weighting and scaling of variables

It may seem intuitively reasonable that some characteristics are of more importance than others in determining the similarity relationships that are present within a data set. For example, two randomly selected molecules are each much more likely to contain a benzenoid ring than to contain a thioether grouping or two gallium atoms, and it has been suggested that such statistically-based information should be used as the basis for weighting schemes in which certain variables are accorded greater or lesser importance than others.

Several types of statistical information can be extracted readily from machine-readable representations of chemical structures to form the basis for a fragment weighting scheme. The simplest such scheme involves a consideration of the number of occurrences of each particular fragment type within a compound, with frequently occurring fragments being given a greater weight than those that occur less frequently. Thus a pair of molecules that had several occurrences of a given fragment in common would be considered as more similar to each other than if they had only a single occurrence in common. The second type of weighting takes into account the sizes of the molecules that are being compared, where we shall assume that the size of a molecule is defined by the number of fragments assigned to that compound. Thus a fragment in a small molecule would be assigned a greater weight than the same fragment in a larger compound, and fragments common to a pair of small molecules would be regarded as a better indication of structural similarity than the sharing of a common fragment by two large structures. The third type of weighting involves the frequency of occurrence of a fragment in the data set, with infrequently occurring fragments being assigned greater weights than those that occur throughout the data set. Thus a pair of compounds that both possessed some rare substructural feature would be regarded as being much more strongly related to each other than if they both possessed a feature such as a benzenoid ring. A further scheme, which has been used in the substructural analysis studies described in Section 1.4.3, takes into account the differential occurrences of substructures in the active and inactive compounds within a data set. Such schemes are of importance in pattern recognition systems which attempt to classify compounds in a training set into active or inactive classes. However, they are of lesser importance in cluster analysis, where the necessary class information is not available and where the main object of the processing is to determine what classes are actually present. Accordingly, this last type of weighting scheme is not considered further here

It should be noted that the weighting of variables in similarity and clustering studies is the subject of some controversy. In particular, Sneath and Sokal (1973) argue strongly against the use of any weighting scheme and list no less than seven reasons in favour of the equal weighting of all of the variables in a study. Their comments relate to taxonomic studies in the life sciences, and some of their points may be less applicable to other subject domains. For example, they note that the use of very many variables results in an averaging out of the effective weight which each variable contributes to the overall measure of similarity; however, this need not be the case if a highly disparate set of weights is used, as with a weighting scheme based upon substructural occurrence data which is known to follow a highly skewed distribution (Adamson et al., 1973). Again, Sneath and Sokal suggest that some precise theory needs to be available so that exact rules are available for calculation of the weights: such theories are available in certain cases (Harrison, 1968; Robertson and Sparck Jones, 1976; Croft and Harper, 1979). Accordingly, a series of experiments has been undertaken to determine the utility of weighting in a chemical context; this work is discussed in Section 3.2.

A less controversial area is that of standardisation, where a mathematical transformation is applied to each of the variables to ensure that they all contribute equally to the determination of similarity. However, there are disagreements even here, as is evidenced in the influential review by Cormack (1971). To see why standardisation might be of importance, consider a data set characterised not only by structural parameters, such as the frequencies of occurrence of particular augmented atom fragments, but also by physicochemical parameters, such as partition coefficients or molar refractivities. The magnitude and the variances of the molar refractivities will typically be very much greater than for the other two types of variable, and the observed inter-molecular similarity relationships will be accordingly determined in large part upon the basis of the refractivities alone, and without regard for the other two types of feature.

A simple way to equalise the magnitudes of different variables is to divide each value by the largest observed value for that variable so that X is replaced by X/XMAX for some variable X (Cain and Harrison, 1958). XMAX can be either the largest observed value for X in the data set or, in some cases, the largest possible value for the variable. An alternative procedure, and one that also attempts to equalise the spread of values in the variables, is that of ranging (Gower, 1971) in which the smallest value for a variable is subtracted from each value and the result divided by the range so that X is replaced by (X-XMIN)/(XMAX-XMIN). However, by far the most common transformation is that of standardisation in which all of the variables are recalculated so as to have zero means and unit variances. This may be achieved by calculating the mean, MX, and standard deviation, SDX, for each variable, and then expressing each observed value in terms of standard deviations from the mean so that X is replaced by (X-MX)/SDX. While there are obvious advantages to be gained from the use of such an approach when different types of variable are available for the characterisation of the objects, it is not at all clear that standardisation should be adopted when all of the variables are of the same type or when there is only a limited number of possible values. An obvious example of such a situation is provided by chemical compounds characterised by substructural fragment occurrence data, where the number of occurrences of a particular fragment type within any structure is likely to be limited to a few integer values. In addition, Sneath and Sokal (1973) note that standardisation is not normally applied to binary data, e.g., fragment bit strings.

Experiments to test the utility of standardisation in chemical applications of cluster analysis are described in Section 4.3.2.1.

2.2.2 Similarity coefficients

Following Sneath and Sokal (1973), four primary classes of similarity coefficient may be identified: distance coefficients, association coefficients, correlation coefficients and probabilistic coefficients. There follow several examples of such coefficients. The discussion assumes that a data set of N compounds is characterised by a total of M different types of descriptor, i.e., chemical fragments, and that the objects, i.e., chemical molecules, are stored for analysis as an NxM matrix, D, where D[I,J] corresponds to the weight of the J'th feature in the I'th object.

Distance coefficients are perhaps the most widely used class of similarity measures owing to their simple geometric interpretation. The most common is probably the Euclidean distance, which, for some pair of objects K and L, is defined to be

$$\left[\sum (D[K,J]-D[L,J])^2\right]^{1/2}$$

The summation in this expression is from J=1 to J=M: this also applies to all of the other expressions in this section. A simple variant of this is the average distance, which is simply the Euclidean distance divided by the number of dimensions, i.e., by M in this case. The Euclidean distance is a special case of a class of metric distance functions called the Minkowski metrics, whose general formula is

$$\left[\sum \left|D[K,J]-D[L,J]\right|^R\right]^{\frac{1}{R}}$$

R=1 corresponds to the Manhattan, or city block, distance and R=2 to the Euclidean distance. A discussion of the Manhattan distance is given by Lance and Williams (1967a) who suggest the use of the so-called Canberra metric

$$\frac{\sum \left|D[K,J]-D[L,J]\right|}{\sum (D[K,J]+D[L,J])}$$

A rather different sort of distance, <u>string distance</u>, may be used to determine the degree of dissimilarity between linear strings of characters. Given a pair of strings, one may be transformed into the other by a series of character insertions, deletions and substitutions, and a quantitative measure of the distance between the two strings may be obtained by assigning a cost function to each of these types of character operation. Such distances are used in specific application contexts such as the comparison of protein and nucleotide sequences for homology and for automatic spelling correction; a detailed discussion of string distances is given by Sankoff and Kruskal (1983). Other types of distance measure are described by Sneath and Sokal (1973).

Association coefficients are used almost as widely as distance measures. Almost all of these coefficients are intended for use with simple two-state, i.e., binary, variables which are conventionally coded as 0 or 1 depending upon their presence or absence within an object description. Although these coefficients can be described in terms of the data matrix, D, it is conceptually simpler to formulate the coefficients as follows for two objects K and L. Let EK0 (EL0) be the number of attributes occurring in K (L) but not in L (K), EKL be the number of attributes in common, and E00 the number of attributes in neither K nor L. Let M be the total number of attributes, i.e. EKL+EK0+EL0+E00. The simplest association coefficient, the <u>simple matching coefficient</u>, is given by (EKL+E00)/(EKL+EK0+EL0+E00). A characteristic of this coefficient, and a problem in the case of chemical structure applications, is that equal importance is given to those features which are present in both structures as to those that are present in neither. In a chemical context, EKL is likely to be considerably less than M while E00 is likely to be broadly comparable with M, and this means that the use of the simple matching coefficient results in a measure of similarity that is dominated by the features absent from both molecules. This does not seem to be a particularly useful definition of similarity, and rather better results may be expected from the use of coefficients that are not so dependent upon the presence of the E00 factor in the numerator. Two commonly used examples of such coefficients are the Dice coefficient

2*EKL/(2*EKL+EKO+ELO) and the Tanimoto coefficient EKL/(EKL+EKO+ELO). Other examples of such coefficients are described by Salton and McGill (1983). The coefficients may readily be generalised to non-binary data, as would be required, for instance, if the data matrix D contained the actual frequencies of occurrence of each fragment type, rather than their mere presence or absence. As an example, the Tanimoto coefficient may be calculated as

$$\frac{\sum D[K,J]*D[L,J]}{\sum (D[K,J]^2+D[L,J]^2-D[J,K]*D[L,J])}$$

Related to this group of coefficients is the <u>cosine coefficient</u>, so called because it measures the cosine of the angle between the K'th and L'th vectors. This coefficient has been used extensively for information retrieval system (Salton and McGill, 1983), and is defined as

$$\frac{\sum D[K,J]*D[L,J]}{\left[\sum D[K,J]^2*\sum D[L,J]^2\right]^{1/2}}$$

Another widely used coefficient of similarity in cluster analysis has been the Pearson product-moment correlation coefficient. Given two structures K and L, let MK be the mean value for all of the variables in the K'th compound (and similarly for ML). Define DIFFKJ as (D[K,J]-MK) and DIFFLJ as (D[L,J]-ML), then the coefficient is defined as

$$\frac{\sum DIFFKJ*DIFFLJ}{\left[\sum DIFFKJ^2*\sum DIFFLJ^2\right]^{1/2}}$$

Although the Pearson coefficient is widely used in general statistical analyses of the relationships between variables, its use for clustering applications has been criticised on the grounds that the terms MK and ML are difficult to interpret, and that a linear relationship between two sets of variables need not be equivalent to a similarity between their parent objects (Eades, 1965; Minkoff, 1965; Gordon, 1981).

The final class of coefficients identified by Sokal and Sneath, probabilistic coefficients, take account of the distribution of the frequencies of variables over the members of a data set. An example of such a coefficient is that described by Goodall (1966). They will not be discussed further here since Adamson and Bush (1975) found that such methods performed poorly with chemical structural data, as well as requiring extremely extensive computations. Moreover, the basic idea underlying such coefficients, that a match on an infrequently occurring variable is of more importance than a match on a common one, may be encompassed by the use of an appropriate weighting scheme as described in the previous section.

There have been a few reports in the literature which describe systematic comparisons of different similarity coefficients: examples are provided by Williams and Dale (1965), Cheetham and Hazel (1969), Green and Rao (1969), Sager and Lockemann (1976), Noreault et al. (1981) and Willett (1983). Such investigations suggest that the similarity coefficient may affect the clustering that is obtained, but there seems to be little consensus as to which types of coefficient are most generally applicable. Sneath and Sokal suggest that one should generally use the simplest type of coefficient that seems appropriate, and point out that such coefficients are often monotonic with more complex measures. A comparison of weighting schemes and similarity coefficients for the measurement of structural similarity is described in Section 3.2.

2.3 HIERARCHIC CLUSTERING METHODS

2.3.1 Introduction

Hierarchic clustering methods result in binary tree-like classifications in which small clusters of objects that are judged to be close, or strongly similar, to each other are nested within larger and larger clusters containing objects that are further separated, or less similar (depending upon whether a similarity or dissimilarity measure is being used). The single cluster containing the entire collection is represented by the root of the tree, while the individual objects reside in the leaves; nodes in the body of the tree correspond to clusters that are formed during the operation of the clustering procedure.

Two main strategies are available for the construction of hierarchic classifications. An agglomerative clustering strategy proceeds through a total of N-1 fusions for a collection of N objects and results in the classification being built upwards from the leaves, with the smallest clusters being generated first and with the final fusion resulting in the root of the tree. Alternatively, a divisive clustering strategy may be adopted in which a single initial cluster is subdivided into smaller and smaller groups of objects.

Hierarchic methods are by far the most popular type of clustering procedure. The methods were developed primarily for applications in the life sciences, where the hierarchic classifications could be compared with traditional biological taxonomies in which specimens are grouped into species, and species grouped into genera. The production of such taxonomies was the first major application of cluster analytic techniques, and it is thus not surprising that the hierarchic methods established a supremacy over the non-hierarchic methods that has remained even when quite different sorts of application area are considered. Extensive discussions of taxonomic applications are given by Sneath and Sokal (1973) and by Clifford and Stephenson (1975), while good general descriptions of hierarchic clustering are provided by Cormack (1971) and Dubes and Jain (1980).

2.3.2 Agglomerative clustering

The most famous of the agglomerative clustering methods is single linkage, or nearest neighbour, in which clusters are formed on the basis of the similarity between the most similar pair of objects, one of which is in each of a pair of clusters. The clusters formed by this method have the property that any cluster member is more similar to at least one member of that cluster than it is to any member of any other cluster; hence the name nearest neighbour. In graph theoretic terms, the single linkage clusters at some similarity level are the connected components of a graph. A characteristic of this method is its tendency to form loosely bound clusters with little internal cohesion, a phenomenon which is referred to as chaining and which has resulted in several attempts to produce related methods that do not suffer from this defect (Shepherd and Willmot, 1968; Wishart, 1969).

An alternative formulation (Gordon, 1981) of the single linkage method assumes that an NxN inter-object distance matrix is available where the KL'th element, DIST[K,L], contains the distances between two objects K and L. Then K and L are defined as belonging to the same cluster at some distance threshold T if there exists a chain of intermediate objects:

$$DIST[K,X_1],\ DIST[X_1,X_2]...DIST[X_N,L]$$

linking them together where each distance is less than T. Thus the number of objects in the cluster will increase as T is increased, with the objects in the cluster at some threshold being a sub-set of those in the cluster at some greater threshold.

It should be noted here that the single linkage method has been described in terms of the characteristics of the clusters that arise from the use of the method; an alternative and popular approach to the characterisation of clustering methods is in terms of the algorithm that gives rise to them. However, there are often many different algorithms available for the implementation of a single clustering method, and the algorithms may differ substantially in their processing requirements: thus Rohlf (1982) describes single linkage algorithms tht have expected time complexities that range between $O(N^2)$ and $O(N^5)$.

A simple single linkage algorithm is as follows:

```
FOR I := 1 TO N-1 DO
  FOR J := I+1 TO N DO
    calculate DIST[I,J] ;
sort the distances into increasing order ;
put each object into a cluster on its own ;
REPEAT
  take the next largest distance, DIST[K,L] ;
  IF K and L are not already in the same cluster THEN
    merge the clusters containing K and L
UNTIL there is only a single cluster.
```

This algorithm takes advantage of the symmetric nature of the inter-object distance matrix, which implies that only N(N-1)/2 of the distances are distinct and need to be calculated and sorted. The distance matrix is of size

$O(N^2)$ and this must be sorted giving overall time and space requirements of at least $O(N^2 \log N^2)$ and $O(N^2)$ respectively. The most efficient current clustering algorithm for the single linkage method is probably that due to Sibson (1973). This SLINK algorithm has time and storage requirements of $O(N^2)$ and $O(N)$ respectively, and involves the progressive updating of the hierarchy as the distances are evaluated.

Closely related to single linkage clustering is the concept of a <u>minimal spanning tree</u>, or MST. Given the set of $N(N-1)/2$ inter-object distances, a spanning tree is a set of $N-1$ distances that links all of the N objects in the data set together into a connected graph without any circuits; the MST is that spanning tree for which the sum of the $N-1$ distances is a minimum, or a maximum if a similarity measure is being used (Lee, 1981). Gower and Ross (1969) showed that the single linkage clusters at some distance threshold T could be obtained by deleting all coefficients from the MST for which the distance was greater than T; a more efficient single linkage algorithm based upon the use of an MST is described by Rohlf (1973), while Zahn (1971) has described a range of clustering procedures based upon the MST. The most generally useful algorithm for the identification of the MST is probably that due to Whitney (1972) while Rohlf (1978) and Bentley and Friedman (1978) have described MST algorithms that may be faster in operation when applied to data sets of low dimensionality; this work is closely related to some of the fast nearest neighbour searching algorithms described in Chapter 5.

Single linkage is just one of a group of hierarchic agglomerative clustering methods that have been used extensively over the years, and which include the complete linkage, group average and Ward methods. All of these methods may be described by the following general algorithm (Lance and Williams, 1967b; Dubes and Jain, 1980; Murtagh, 1983):

```
FOR I := 1 TO N-1 DO
  FOR J := I+1 TO N DO
    calculate DIST[I,J] ;
REPEAT
  search DIST to identify the closest remaining
   pair of clusters, K and L ;
  fuse K and L to form a new cluster KL ;
  update DIST by calculating the distance from KL
    to all remaining clusters
UNTIL there is only a single cluster.
```

DIST contains $O(N^2)$ elements and is searched N-1 times, once for each fusion; accordingly the algorithm has overall time and space requirements of $O(N^3)$ and $O(N^2)$ respectively. The various hierarchic agglomerative methods differ in the definition of distance used for the selection of the closest pairs of clusters and for the updating of DIST.

The complete linkage, or furthest neighbour, method is the converse of single linkage, since the least close pair of objects forms the basis for the definition of inter-cluster distance; thus, each cluster member is closer to the furthest member of that cluster than to the furthest member of any other cluster. This definition of cluster membership is very much stricter than that for single linkage, and thus the large straggly clusters in the latter case are here replaced by large numbers of small tightly bound groupings, a form of classification which may be just as inappropriate as the extended single linkage clusters in some cases. In graph theoretical terms, complete linkage clustering corresponds to the identification of the maximally complete subgraphs at some threshold distance.

The group average method results in clusters such that each cluster member has a lesser average distance to the remaining members of that cluster than it does to all members of any other cluster. It thus represents a midpoint between the two extremes represented by single linkage and complete linkage but has been criticised for tending to produce small 'non-conformist' clusters of outliers that are unlike every other cluster (Edelbrock, 1979). There are several other average linkage methods, such as the median and centroid methods (Sneath and Sokal, 1973), but these suffer from the fact that inversions may arise in the dendrogram representing a hierarchy.

Ward's method fuses those two clusters whose fusion results in the least increase in the sum of the distances from each document to the centroid of its cluster. Ward's method has proved to be an extremely powerful grouping mechanism, and Wishart (1978) suggests that it is probably the most generally useful hierarchic procedure. It has, however, been criticised for tending to produce spherical clusters which may not accurately reflect the true shape of the clusters present in the data set.

The general algorithm above has two useful features. Firstly, the entries in the distance matrix corresponding to the new cluster KL can over-write those corresponding to the predecessor clusters K and L; thus, once the initial distances have been calculated no additional storage is required during the classification. The second point is that the new entries in the matrix corresponding to the distances between KL and each of the other clusters may be calculated from the entries for K and L without recourse to the original data matrix. This may be done using a recurrence formula due to Lance and Williams (1967b). As with single linkage, a wide range of alternative algorithms is available for the implementation of the other hierarchic agglomerative methods; recent work in this area is described by Murtagh (1983, 1984).

2.3.3 Divisive clustering

As noted previously, the agglomerative clustering methods are not the only means by which a hierarchic classification of a data set may be obtained. Whereas the agglomerative methods proceed by a series of fusions so that, after the final fusion, all of the objects are in a single cluster, divisive clustering methods are based upon the progressive sub-division of an initial single cluster into smaller and smaller sub-clusters. Edwards and Cavalli-Sforza (1965) have shown that there are $2^{N-1}-1$ different ways in which a group of N objects can be sub-divided into 2 non-empty sub-groups, and means must hence be found to restrict the number of possible divisions. Polythetic divisive methods have been described in the literature (MacNaughton-Smith et al., 1964)), but the monothetic divisive methods (Clifford and Stephenson, 1975; Gordon, 1981) have been used much more extensively, and some of them will be discussed briefly here. These methods are based on the division of each cluster into two sub-clusters, membership of which is dependent upon the presence or absence of some chosen feature; all of the objects in one of the clusters resulting from a division will contain that attribute, while it will be absent from the attribute vectors of the objects in the other new cluster that has been formed. A good general discussion of divisive clustering is given by Wishart (1978).

Clifford and Stephenson (1975) suggest that monothetic divisive methods have several advantages over the more common agglomerative procedures. Firstly, the group definitions are simple and unambiguous, resulting as they do from the absence or presence of a single attribute. Secondly, divisive methods should, in theory, be superior, because the initial splits take place on the entire data set where the maximum amount of information about attribute frequencies and co-occurrences is available. In contrast, agglomerative methods start with individual pairs of objects, and thus the final classification may be strongly dependent upon the characteristics of individual members of the data set. Clifford and Stephenson also suggest that the computational requirements for divisive clustering may be less than for agglomerative clustering, although still being extensive in some cases. An efficient algorithm for large scale monothetic divisive clustering is described by Openshaw (1980).

The attribute that is chosen as the basis for the division at each stage in the clustering will be that which maximises the dissimilarity between the new clusters that are formed. The four methods described here, and tested experimentally in Section 4.3.3, differ in the criteria used for the selection of the discriminating attribute.

The earliest approach to monothetic divisive clustering was association analysis (Williams and Lambert, 1960), which has been used extensively in ecological applications of cluster analysis. It involves the selection of that attribute at a given stage in the classification process which is most strongly associated with the rest of the attributes, the strength of the association being determined by a modified chi squared test. Specifically, for a data set containing M attributes, the chi squared values are calculated for all M(M-1)/2 pairs of attributes; these values are summed over all attributes occurring with each specified one, and that with the largest sum used as the basis for the division of the set of objects into those objects that contain the attribute, and those objects that do not.

An alternative chi squared-like measure, called the interaction statistic, forms the basis for the group analysis method of Crawford and Wishart (1967). An advantage of the procedure is that the individual attribute scores here may be calculated from a simple serial scan of the data matrix, thus making the procedure very much more efficient in operation.

Lance and Williams (1968) describe a method, called information analysis, which selects at each stage in the division that attribute which maximises the change in the so-called information content of a cluster when sub-divided into two, information content being a function similar to that used in information theory to measure the capacity of a communication channel.

Finally, the error sum of squares method involves the selection of that attribute which minimises the separation between each of the objects and the centroid of its cluster; a similar criterion forms the basis for the Ward agglomerative clustering method and the non-hierarchic partitioning methods described in the next section.

The major limitation of the monothetic divisive methods is that an object can be more similar to objects belonging to a different group than to members of its own cluster, depending upon which variables happened to be chosen as a basis for the division of the data set (Gordon, 1981). For this reason, monothetic division is sometimes followed by a subsequent relocation procedure (Crawford and Wishart, 1968; Weir, 1970) similar to those described in Section 2.4.2.

2.4 NON-HIERARCHIC CLUSTERING METHODS

2.4.1 Introduction

Non-hierarchic clustering methods divide a data set into a series of sub-sets, with similar objects in the same cluster being separated from non-similar objects which are placed in different clusters. Such a partition describes a classification in which there are no hierarchic relationships between the various clusters that have been identified by the clustering procedure.

An optimal solution to the problem of finding a partition may be obtained by the simple expedient of defining some criterion of the 'goodness' of a classification, and then systematically generating and evaluating all possible partitions to see which of them best satisfies this criterion. The problem here is that a complete enumeration is quite infeasible for data sets of anything but trivial size; for example, there are about 193 million possible partitions of 19 objects into 3 clusters (Duran and Odell, 1974). Because of this, a range of heuristics has been described which permit a sub-optimal partition to be obtained at a much lower computational cost.

The most common approach takes the error sum of squares as the basic clustering criterion, the aim being to partition the set of N objects into C clusters so as to minimise the total within-cluster sum of squares about the C cluster centroids. This simple and intuitively appealing idea has been met above in the Ward agglomerative and error sum of squares divisive methods, and can be analysed on a formal basis using statistical methodologies (Gordon, 1981). It forms the basis for a large number of non-hierarchic methods that differ only slightly in the exact manner in which they are implemented (Aldenderfer, 1977; Dubes and Jain, 1980; Spath, 1980) and which are described in the next section.

2.4.2 Relocation clustering

A simple partitioning algorithm is that described by Forgy (1965), as shown below, where C is the requisite number of clusters:

```
FOR I := 1 TO C DO
  BEGIN
    select an object at random ;
    the I'th cluster centroid := this object
  END ;
REPEAT
  FOR I := 1 TO N DO
    assign the I'th object to the closest centroid ;
  FOR I := 1 TO C DO
    calculate all of the resulting centroids
UNTIL the classification has stabilised.
```

The repeated assignments result in objects being relocated from one cluster to another as the classification proceeds, and this has resulted in such methods being called <u>relocation</u> clustering. It will be clear that small modifications of this outline algorithm permit a large number of possible methods depending upon the exact manner in which it is implemented. Thus, various means are possible for specifying the number of initial clusters, the cluster centroids, the manner in which these centroids are recomputed, and the termination condition.

The selection of C is perhaps the most complicated problem here, and one that is by no means solved (Ling, 1971; Sneath and Sokal, 1973). An example of work in this area is the study of Beale (1969) who describes a statistic which may be used to test whether a partition into some number of clusters is significantly better than an alternative sub-division into a smaller number of clusters. Other approaches are reviewed by Everitt (1979), but the most widely used procedure seems to be the investigator's estimate of how many distinct groups are present within the data set. It may be noted that entirely comparable problems are encountered if a partition is obtained by applying some similarity threshold to a hierarchic classification, so as to identify the clusters at that similarity, since then some criterion must be used to decide the threshold (Baker and Hubert, 1975; Mojena, 1977).

The use of random selection implies that different final partitions may be obtained from different initial assignments, and several means have thus been suggested for non-random assignments that would eliminate this arbitrary characteristic of relocation clustering (Dattola, 1971; Hartigan and Wong, 1977; Willett, 1980; Can and Ozkarahan, 1984). Milligan (1980) suggested that relocation clustering performs best if the initial centroids were generated by a group average clustering of the data set.

The centroid definition used is generally the arithmetic mean of the attribute vectors of all of the objects currently within a given cluster. However, document clustering applications in information retrieval (Van Rijsbergen, 1979) have often adopted a simpler procedure. Here, each of the documents is characterised by the presence or absence of some small number of keyword descriptors that have been selected from a much larger indexing vocabulary, and a document-like centroid may be obtained by selecting those keywords that occur in more than some threshold number of documents in that cluster. Both approaches are used in the chemical structure clustering experiments described in Chapter 4.

The Forgy algorithm involves the recomputation of the cluster centroid at the end of a pass, i.e., when all of the objects have been assigned to a cluster. An alternative procedure involves recomputing the centroids immediately after an object changes its cluster membership. This may increase the speed of convergence, and has also been found to result in superior classifications in Monte Carlo simulation experiments (Milligan, 1980); it does, however, mean that the final classification will again be dependent upon the order in which the data set is processed. This is a rather undesirable characteristic of many non-hierarchical clustering methods.

Two types of procedure are commonly used for the assignment of objects to clusters during the course of the relocation. The K-means approach involves the assignment of each object to its closest centroid, whereas a hill-climbing assignment procedure may be used in which an object changes its membership only if the change results in an improvement in the chosen criterion function. This latter approach is adopted in the CLUSTAN package which was used for the relocation experiments reported in Section 4.3.4.

Finally, some condition must be specified to ensure that the relocation does not continue indefinitely. It is possible to show that certain types of relocation procedure are guaranteed to terminate, i.e., to reach a stage at which no object changes its assignment during a single iteration of the algorithm (Dattola, 1971), but in practice this may not happen very rapidly. Accordingly, relocation algorithms normally terminate when no more than some small number of the objects change their assignment during a single iteration, when a statistical measure-of-goodness reaches some threshold value, or after some fixed number of iterations. In practice, it has often been found that only a few objects change their cluster membership once a relocation algorithm has been iterated two or three times.

An example of a more sophisticated relocation method is the well-known ISODATA procedure (Ball and Hall, 1967) which adopts a complex initialisation routine, and allows the use of user-defined parameters to control the elimination or merger of very small clusters, and the splitting of clusters which have grown to be too large; good overviews of the ISODATA algorithm are given by Tou and Gonzalez (1974) and by Dubes and Jain (1980).

An inspection of the Forgy algorithm shows that the time complexity is of order O(CN), where typical values for C include $N^{1/2}$ and $\log_2 N$. It is thus clear that the computational requirements here are considerably less than the time complexities of $O(N^2)$ or greater associated with clustering methods based on an inter-object similarity matrix. Relocation clustering is thus fast in operation and appropriate to the clustering of large data sets.

Comparisons of different squared-error relocation

methods (Dubes and Jain, 1976; Aldenderfer, 1977) suggest that they often give similar partitions despite their very disparate modes of operation, suggesting that a simple and computationally efficient algorithm is likely to be as useful as a more complex one. FORTRAN listings for relocation clustering are given by Hartigan (1975) and by Spath (1980).

2.4.3 Single pass clustering

A very simple type of squared-error procedure is the single pass algorithm in which, as the name implies, the repeated assignments of a Forgy-like algorithm are replaced by a single pass through the data. The basic algorithm is shown below, with T being a user-defined threshold distance, and with C here being the number of clusters identified at any point during the classification.

```
C := 0 ;
FOR I := 1 TO N DO
  BEGIN
    compare the I'th object with the C clusters
      to identify the closest distance ;
    IF this distance < T THEN
      BEGIN
        the object joins the corresponding cluster;
        the centroid is recomputed
      END
    ELSE
      BEGIN
        C := C+1 ;
        the object becomes the C'th cluster
      END
  END.
```

This algorithm is much faster than the relocation methods

but suffers from being totally dependent upon the order in which the objects are processed, with clusters identified early on having the opportunity to become very much larger than those identified near to the end of the file. A variety of procedures have been suggested to 'tidy-up' the classifications produced by such an algorithm (Fritsche, 1974), and the methods have been used quite extensively for information retrieval applications (Salton, 1971), where very large data sets need to be handled, and as a means of specifying the initial centroids for input to a relocation method (Hartigan, 1975).

2.4.4 Near neighbour and other non-hierarchic methods

Another group of non-hierarchic clustering methods derives from the work of Jarvis and Patrick (1973) and involves the use of a near neighbour table. This is an NxK integer array that contains the identifiers for the K nearest neighbours for each of the N objects, with the IJ'th element containing the J'th nearest neighbour for the I'th object. Jarvis and Patrick suggest that two objects, P and Q, should be placed in the same cluster if P is a near neighbour of Q, Q is a near neighbour of P, and P and Q share at least R near neighbours in common, where R is a user-defined parameter. Progressive incrementing of R allows the production of a hierarchic sequence of partitions (Gowda and Krishna, 1978a; Gowda and Krishna, 1978b; Mizoguchi and Shimura, 1980). Such methods are simple in concept and efficient in operation, since the nearest neighbours may be identified using the fast search algorithms described in Chapter 5; the computational requirements are thus between those of the relocation and hierarchic agglomerative methods. Jarvis-Patrick clustering is considered in detail in Sections 4.3.5 and 4.4.

Other non-hierarchic methods that have been described include density estimation, or mode seeking, algorithms, which try to identify local peaks, or modes, in the density function describing the data set that could be used as cluster centres, mixture resolution techniques, which assume that the objects are samples from a finite mixture of multi-normal distributions and which try to estimate the parameters of such distributions, and fuzzy clustering, which allows objects to belong to more than one cluster in the partition, with different measure of 'belongingness' for each cluster. The reviews of Everitt (1974) and of Dubes and Jain (1980) provide an opening into this work.

2.5 THE COMPARISON OF CLUSTERING METHODS

With such a wide range of methods available, an investigator who wishes to use cluster analytic techniques needs guidelines as to which methods should be used. There has hence been a considerable amount of interest in methods for the evaluation and comparison of different clustering methods. Three main classes of evaluation may be noted, these being theoretical analyses of the individual methods, simulation studies of the classifications resulting from their use, and purely empirical comparisons of their utility in particular environments. There follows a brief discussion of these evaluation methods; comprehensive reviews are given by Dubes and Jain (1976, 1979, 1980), Rohlf (1974) and Raghaven and Ip (1982).

Theoretical studies (Gower, 1967; Jardine and Sibson, 1968; Jardine and Sibson, 1971; Fisher and van Ness, 1971) attempt to provide a mathematical analysis of the characteristics of various methods. Jardine and Sibson

(1968) proposed several intuitively reasonable conditions that any admissible clustering method should satisfy, including the requirements that a unique result should be obtained from a given data set, and that the classification should be unaffected by monotonic transformations of the input similarity matrix and by small changes in the basic data. They found that, of the hierarchic methods studied, only single linkage satisfied all of the conditions. A similar investigation, but with a quite different set of criteria, was carried out by Fisher and van Ness (1971), who again found that single linkage was the most acceptable of the methods tested. However, the conclusion that single linkage should be the method of choice has been strongly criticised by Williams et al. (1971) who feel that the suggested criteria are too stringent and that other methods are to be preferred owing to their generally better performance in practical applications. A discussion of these opposing views is given by Aldenderfer (1977).

Simulation studies involve the generation of an artificial data set so that the true structure of the data is known. The performance of a clustering method may then be assessed by determining the degree to which it can recover the true structure using an index of recovery such as the Rand (1971) or kappa (Cohen, 1960) statistics. Experiments have been reported to identify the basic recovery characteristics of methods and their stability when various error conditions are used to perturb the original data (Kuiper and Fisher, 1975; Blashfield, 1976; Mojena, 1977; Bayne et al., 1980; Milligan and Isaac, 1980; Raghaven and Yu, 1981). A comprehensive review of the techniques and results of several such Monte Carlo tests is given by Milligan (1981), who states that the Ward and group average methods tend to give the best recovery performance while the single linkage method performs consistently poorly. However there are severe methodological problems associated with the analyses, and the experiments to date have not been entirely in agreement.

Because of the difficulties associated with these two types of comparison, Everitt (1979) has called for more practical studies on real data sets which try to evaluate clustering methods on the basis of the usefulness of the classifications produced. Such comparative tests have been carried out in a range of very different application areas. Thus Dubes and Jain studied clusterings of handwritten characters (1976) and choral speech records (1979), Mezzich (1978) compared automatic and manual classifications of psychiatric patients with an existing diagnostic classification, Morey et al. (1983) evaluated hierarchic and non-hierarchic classifications of alcohol abusers on the basis of a range of socio-economic factors, while Griffiths et al. (1984, 1986) and Vorhees (1985) compared agglomerative methods by their effectiveness for information retrieval. There is little agreement from such comparative tests about the relative merits of the different clustering methods, although the single linkage method often seems to perform less well than other methods. Mezzich (1978) found that the best results were obtained with the K-means or complete linkage methods, while the study of relocation methods by Dubes and Jain (1976) suggested that simple procedures were often comparable in performance with more sophisticated ones. Griffiths et al. (1984) and Vorhees (1985) found that single linkage was noticeably inferior to the other methods that were tested, these all being of comparable effectiveness for document retrieval, but a later study suggested that Ward's method gave the best retrieval performance (Griffiths et al., 1986).

Section 4.3 describes an extended comparison of a whole range of clustering methods for the classification of chemical structures.

2.6 THE USE OF SIMILARITY AND CLUSTERING METHODS

Clustering can be used for many purposes, these depending in large part upon the context in which the methods are being used. In the context of chemical information systems, Chapter 1 has identified the areas of structure and substructure searching and QSAR as the most important areas of potential application, and we now consider in outline how similarity and clustering techniques could be most usefully applied.

The selection of a search algorithm for access to a repository of stored data is conditioned by many factors. Thus a choice may often need to be made between the efficiencies of retrieval and of database update as records need to be added, deleted or modified, while considerations of storage or software overheads may also be of importance. However, perhaps the most important factor is the search facilities that are likely to be required, and three major types may be identified.

An exact match search of a data base involves the identification of an exact match between the query statement, however it is expressed, and one or more of the records in the file; an example of this is a registration search as described in Section 1.3.1. Partial match, or inclusive match, searches require the retrieval of those records which satisfy all of the constraints in the query, but which may possess other characteristics as well in specified data fields; such searches characterise many current retrieval systems, whether for numeric, structural or textual data. An obvious example of partial match retrieval is substructure searching, as described in Section 1.3.2, where all of the molecules retrieved in a search contain the query as a subgraph. Best match searching involves the identification of those records in the file which are most similar to the query record; alternative names for this type of search include nearest neighbour and ranked output.

Best match retrieval is characteristic of real-world problems where the data may be imprecise, incomplete or ill-defined in some way; examples of this include text retrieval systems (Salton and McGill, 1983), spelling error correction routines (Peterson, 1980), and classification problems in many pattern recognition systems (Tou and Gonzales, 1974). Such a search mechanism could provide a useful adjunct to the search facilities available in current chemical information systems since it would permit the ranking of compounds in order of decreasing similarity to a query structure or substructure, the ranking being based upon some quantitative measure of inter-molecular similarity or distance. A ranking search would overcome many of the problems that are associated with current searching systems (Salton and McGill, 1983). Many of these advantages relate to the Boolean searching that characterises substructural retrieval but, as will be shown in Section 3.3.3, there are also potential advantages in using ranking methods to allow a quite novel type of structure searching.

The first main disadvantage of the Boolean model is the complete lack of control over the size of the output that is produced by a particular query. Unless the chemist has a detailed knowledge of the contents of the file, (s)he will be quite unable to predict a priori how many compounds will be identified that contain the query; it may be several hundreds if the query has been phrased in very general terms, or it may be none at all if too detailed a substructure has been input. In both cases, the chemist will need to reformulate the query in some way and then to carry out a second search that will, hopefully, retrieve a more useful number of structures. If best match searching is available, the chemist could choose to see as much, or as little, of the output as required; moreover, whatever number of structures is inspected, the chemist knows that these are the molecules that the system judges to be most similar to the query structure or substructure.

The second limitation of Boolean searching is that the retrieval algorithm results in a simple partition of the data base into two discrete sub-sets: those molecules that satisfy the structural constraints of the query and those that do not. This binary partition has two major consequences. It implies, firstly, that the chemist carrying out the search must define the query structure as explicitly as possible, specifying, e.g., the positions and the types of all of the functional groups that are acceptable. The second consequence is that all of the retrieved compounds are presumed to be of equal usefulness to the searcher, and there is no mechanism by which they can be ranked in order of decreasing probability of utility. Such problems are eliminated when the retrieval mechanism automatically provides a ranking of the output.

Finally, the Boolean model does not allow the system or users to reflect the relative importance of different components of the query. As has been described above in

Sections 1.4.3 and 2.2.1, several means are available for the calculation of weights reflecting the relative importance of different fragments in determining the overall degree of similarity between a pair of objects. However, whereas fragment weighting may form an important part of any retrieval algorithm based on structural similarity, Boolean searching implicitly assumes that all of the fragments have weights of either 1 or 0, depending upon whether they happen to be present or absent in the query.

It would thus seem that the identification of some suitable measure of inter-molecular similarity could be used as the basis for a best match chemical search system offering facilities additional to those available in current chemical information systems. This conclusion is studied in detail in Chapters 3 and 4. As will be seen from the examples presented in Section 3.3, the use of best match searching allows a chemist to see a wide range of classes of compound that are structurally related to a chosen target compound. Two types of search facility are presented: the ranking of the output of a conventional substructure search, and the ranking of an entire set of compounds. In both cases, compounds are ranked in order of decreasing similarity to an input target structure of interest to the chemist who is running the search. A further type of best match searching mechanism is discussed in Section 4.4, the aim here being to cluster the output from a substructure search. While a substructural query may be sufficiently discriminating to retrieve just a few compounds that exactly reflect the interests of the searcher, it is often the case that a general query may result in very many molecules being retrieved in response to the search. In such a case, a useful retrieval facility would provide for the grouping of the output so that the chemist could obtain a rapid over-view of the range of structural types present by inspecting just a single molecule from each cluster. Accordingly, this provides a mechanism for identifying classes of compounds and complements the facilities presented in Chapter 3, which permit the identification of a particular class of compounds from amongst the classes present in the file.

The provision of new searching techniques is the main application area considered in this book. However, similarity and clustering methods may also be used as an aid to drug and pesticide development programmes. As noted in Section 1.4, one of the principal assumptions underlying QSAR studies is that structurally similar molecules will exhibit similar properties, and thus an effective best match search system provides an obvious means of identifying compounds with properties comparable with those of the target molecule used in the search. While such a retrieval-based mechanism is very much cruder than even the most heuristic QSAR methods, it does provide a simple and efficient means of identifying sufficient compounds for the application of more sophisticated techniques. As an example, a best match search using a single compound of known activity followed by testing of some of the highly ranked compounds might provide sufficient occurrence data for the calculation of activity and inactivity weights for a substructural analysis study. It should be noted, however, that the resulting sample is, of course, very highly biased towards the original target structure.

Clustering methods may also be used in the initial stages of a drug development programme, when compounds are being chosen for random testing against primary biological screens. The basic idea here is to cluster an entire collection of compounds so as to identify the main classes present in the data. Compounds may then be selected systematically from the clusters to ensure that all types of structure have the chance to be tested for activity in the screen. Again this could hardly be considered as a way of identifying a QSAR, but it does provide a useful facility to support pharmaceutical research. The potential utility of such an approach is considered in Section 4.4.1.

The provision of the searching and drug design facilities discussed in the previous paragraphs requires that the similarity and clustering routines are both effective and efficient in operation: effective in that useful results are obtained, and efficient in that the facilities can be provided at a moderate computational cost. Much of the remainder of this book is devoted to an extended comparison of the effectiveness and efficiency of a very wide range of similarity and clustering techniques. In particular, Section 3.2 compares the effectiveness of different types of similarity measure, and Section 4.3 the effectiveness of different types of clustering method; in both cases, the evaluation is carried out by simulated property prediction. These extended comparative experiments are restricted to small data sets so as to enable the testing of as wide a range of methods as possible. Once an appropriate technique has been identified, it may be used for the processing of large numbers of structures, viz., the searching and clustering experiments of Sections 3.3 and 4.4 respectively. Such large scale processing can be achieved if ,and only if ,an efficient implementation of the technique is available, since it is clearly of little use to identify some technique as the most appropriate only to then find that it cannot be applied to data sets of realistic size. Accordingly, the problem of algorithmic efficiency is considered in some detail in Chapter 5.

REFERENCES

Adamson, G.W. and Bush, J.A. (1975). A comparison of the performance of some similarity and dissimilarity measures in the automatic classification of chemical structures. J. Chem. Inform. Comput. Sci. 15, 55-58.

Adamson, G.W., Cowell, J., Lynch, M.F., McLure, A.H.W., Town, W.G. and Yapp, A.M. (1973). Strategic considerations in the design of a screening system for substructure searches of chemical structure files. J. Chem. Docum. 13, 153-157.

Aldenderfer, M.S. (1977). A Consumer Report On Cluster Analysis Software. 4 vols. Pennsylvania State University, Philadelphia.

Baker, F.B. and Hubert, J.L. (1975). Measuring the power of hierarchical cluster analysis. J. Am. Stat. Assoc. 70, 31-38.

Ball, G.H. and Hall, D.J. (1967). A clustering technique for summarizing multivariate data. Behav. Sci. 12, 153-155.

Bayne, C.K., Beauchamp, J.J., Begovich, C.L. and Kane, V.E. (1980). Monte Carlo comparisons of selected clustering procedures. Patt. Recog. 12, 51-62.

Beale, E.M.L. (1969). Euclidean cluster analysis. Bull. ISI 43, 92-94.

Bentley, J.L. and Friedman, J.H. (1978). Fast algorithms for constructing minimal spanning trees in coordinate spaces. IEEE Trans. Comput. C-27, 97-105.

Blashfield, R.K. (1976). Mixture model tests of cluster analysis: accuracy of four hierarchical methods. Psych. Bull. 83, 377-388.

Bush, J.A. (1977). Automatic Classification and Chemical Structure-Activity Correlation. Ph.D. thesis, University of Sheffield.

Cain, A.J. and Harrison, G.A. (1958). An analysis of the taxonomists' judgement of affinity. Proc. Zool. Soc. Lond. 131, 85-98.

Can, F. and Ozkarahan, E.A. (1984). Two partitioning type clustering algorithms. J. Am. Soc. Inform. Sci. 35, 268-276.

Cheetham, A.H. and Hazel, J.E. (1969). Binary (presence-absence) similarity coefficients. J. Paleont. 43, 1130-1136.

Clifford, H.T. and Stephenson, W. (1975). An Introduction To Numerical Classification. Academic Press, New York.

Cohen, J.A. (1960). A coefficient of agreement for nominal scales. Educ. Psychol. Measur. 20, 37-46.

Cormack, R.M. (1971). A review of classification. J. Roy. Stat. Soc. 134, 321-367.

Crawford, R.M.M. and Wishart, D. (1967). A rapid multivariate method for the detection and classification of groups of ecologically related species. J. Ecol. 55 505-525.

Crawford, R.M.M. and Wishart, D. (1968). A rapid classification and ordination method and its application to vegetation mapping. J. Ecol. 56, 385-404.

Croft, W.B. and Harper, D.J. (1979). Using probabilistic models of document retrieval without relevance information. J. Docum. 35, 285-295.

Dattola, R.T. (1971). Experiments with a fast algorithm for automatic classification. in The SMART Retrieval System. (Edited by G. Salton). Prentice-Hall, Englewood Cliffs.

Dubes, R. and Jain, A.K. (1976). Clustering techniques: the user's dilemna. Patt. Recog. 8, 247-260.

Dubes, R. and Jain, A.K. (1979). Validity studies in

clustering methodologies. Patt. Recog. 11, 235-254.

Dubes, R. and Jain, A.K. (1980). Clustering methodologies in exploratory data analysis. Adv. Comput. 19, 113-228.

Duran, B.S. and Odell, P.L. (1974). Cluster Analysis: A Survey. Springer-Verlag, Berlin.

Eades, D.C. (1965). The inappropriateness of the correlation coefficient as a measure of taxonomic resemblance. Syst. Zool. 14, 98-100.

Edelbrock, C. (1979). Mixture model tests of hierarchical clustering algorithms: the problem of classifying everybody. Mult. Behav. Res. 14, 367-384.

Edwards, A.W.F. and Cavalli-Sforza, L.L. (1965). A method for cluster analysis. Biometrics 21, 362-375.

Everitt, B.S. (1974). Cluster Analysis. Heinemann, London.

Everitt, B.S. (1979). Some unresolved problems in cluster analysis. Biometrics 35, 169-181.

Fisher, L. and van Ness, J.W. (1971). Admissible clustering procedures. Biometrika 58, 91-104.

Forgy, E. (1965). Cluster analysis of multivariate data: efficiency versus interpretability of classifications. Biometrics 21, 768.

Fritsche, M. (1974). Automatic Clustering Techniques In Information Retrieval. Commission of the European Communities, Luxembourg.

Goodall, D.W. (1966). A new similarity index based on probability. Biometrics 22, 882-907.

Gordon, A.D. (1981). Classification. Chapman and Hall, London.

Gowda, K.C. and Krishna, G. (1978a). Agglomerative clustering using the concept of mutual nearest neighbourhood. Patt. Recog. 10, 105-112.

Gowda, K.C. and Krishna, G. (1978b). Disaggregative clustering using the concept of mutual nearest neighbourhood. IEEE Trans. Syst. Man. Cybernet. SMC-8, 888-889.

Gower, J.C. (1967). A comparison of some methods of cluster analysis. Biometrics 23, 623-637.

Gower, J.C. (1971). A general coefficient of similarity and some of its properties. Biometrics 27, 857-871.

Gower, J.C. and Ross, G.J.S. (1969). Minimum spanning trees and single linkage cluster analysis. Appl. Stat. 18, 54-64.

Green, P.E. and Rao, V.R. (1969). A note on proximity measures and cluster analysis. J. Market. Res. 6, 359-364.

Griffiths, A., Robinson, L.A. and Willett, P. (1984). Hierarchic agglomerative clustering methods for automatic document classification. J. Docum. 40, 175-205.

Griffiths, A., Luckhurst, H.C. and Willett, P. (1986). Using nearest neighbour information in document retrieval systems. J. Amer. Soc. Inform. Sci. 37, 3-11.

Harrison, P.J. (1968). A method of cluster analysis and some applications. Appl. Stat. 17, 226-236.

Hartigan, J.A. (1975). Clustering Algorithms. Wiley, New York.

Hartigan, J.A. and Wong, M.A. (1979). A K-means clustering algorithm. Appl. Stat. 28, 100-108.

Jardine, N. and Sibson, R. (1968). The construction of hierarchic and non-hierarchic classifications. Comput. J. 11, 177-184.

Jardine, N. and Sibson, R. (1971). Mathematical Taxonomy. Wiley, New York.

Jarvis, R.A. and Patrick, E.A. (1973). Clustering using a similarity measure based on shared nearest neighbours. IEEE Trans. Comput. C-22, 1025-1034.

Kuiper, F.K. and Fisher, L. (1975). A Monte Carlo comparison of six clustering procedures. Biometrics 31, 777-783.

Lance, G.N. and Williams, W.T. (1967a). Mixed-data

classificatory program. I. Agglomerative systems. Aust. Comput. J. 1, 15-20.

Lance, G.N. and Williams, W.T. (1967b). A general theory of classificatory sorting strategies. I. Hierarchical systems. Comput. J. 9, 373-380.

Lance, G.N. and Williams, W.T. (1968). Note on a new information-statistic classificatory program. Comput. J. 11, 195.

Lee, R.C. T. (1981). Clustering analysis and its applications. Adv. Inform. Syst. Sci. 8, 169-292.

Ling, R.F. (1971). Cluster Analysis. PhD thesis, Yale University.

MacNaughton-Smith, P., Williams, W.T., Dale, M.B. and Mockett, L.G. (1964). Dissimilarity analysis: a new technique of hierarchical sub-division. Nature 202, 1034-1035.

Mezzich, J.E. (1978). Evaluating clustering methods for psychiatric diagnosis. Biol. Psychiat. 13, 265-281.

Milligan, G.W. (1980). An examination of the effect of six types of error perturbation on fifteen clustering algorithms. Psychomet. 45, 325-342.

Milligan, G.W. and Isaac, P.D. (1980). The validation of four ultrametric clustering algorithms. Patt. Recog. 12, 41-50.

Milligan, G.W. (1981). A review of Monte Carlo tests of cluster analysis. Mult. Behav. Res. 16, 379-407.

Minkoff, E.C. (1965). The effects on classification of slight alterations in numerical techniques. Syst. Zool. 14, 196-213.

Mizoguchi, R. and Shimura, M. (1980). A nonparametric algorithm for detecting clusters using hierarchical structure. IEEE Trans. Patt. Anal. Mach. Intell. PAMI-2, 292-300.

Mojena, R. (1977). Hierarchical grouping methods and stopping rules: an evaluation. Comput. J. 20, 359-363.

Morey, L.C., Blashfield, R.K. and Skinner, H.A. (1983). A comparison of cluster analysis techniques within a sequential validation framework. Mult. Behav. Res. 18, 309-329.

Murtagh, F. (1983). A survey of recent advances in hierarchical clustering algorithms. Comput. J. 26, 354-359.

Murtagh, F. (1984). Complexities of hierarchic clustering algorithms: state of the art. Comput. Stat. Quart. 1, 101-114.

Noreault, T., McGill, M. and Koll, M.B. (1981). A performance evaluation of similarity measures, document term weighting schemes and representations in a Boolean environment. in Information Retrieval Research. (Edited by R.N. Oddy, S.E. Robertson, C.J. Van Rijsbergen and P.W. Williams). Butterworth, London.

Openshaw, S. (1980). Monothetic divisive algorithms for classifying large data sets, in COMPSTAT 1980. (Edited by International Association for Statistical Computing). Physica-Verlag, Vienna.

Peterson, J.L. (1980). Computer programs for detecting and correcting spelling errors. Comm. ACM 23, 676-687.

Raghaven, V.V. and Yu, C.T. (1981). A comparison of the stability characteristics of some graph theoretic clustering methods. IEEE Trans. Patt. Anal. Mach. Intell. PAMI-3, 393-402.

Raghaven, V.V. and Ip, M.Y.L. (1982). Techniques for measuring the stability of clustering: a comparative study. Lect. Notes Comput. Sci. 146, 209-237.

Rand, W.M. (1971). Objective criteria for the evaluation of clustering methods. J. Am. Stat. Assoc. 66, 846-850.

Robertson, S.E. and Sparck Jones, K. (1976). Relevance weighting of search terms. J. Am. Soc. Inform. Sci. 27, 129-146.

Rohlf, F.J. (1973). Hierarchical clustering using the

minimum spanning tree. Comput. J. 16, 93-95.

Rohlf, F.J. (1974). Methods of comparing classifications Ann. Rev. Ecol. Systemat. 5, 101-113.

Rohlf, F.J. (1978). A probabilistic minimum spanning tree algorithm. Inform. Proc. Lett. 7, 44-48.

Rohlf, F.J. (1982). Single-link clustering algorithms. Handbook Stat. 2, 267-284.

Sager, W.K.H. and Lockemann, P.C. (1976). Classification of ranking algorithms. Int. Forum Inform. Doc. 1, 12-25.

Salton, G. (1971). The SMART Retrieval System Prentice-Hall, Englewood Cliffs.

Salton, G. and McGill, M.J. (1983). Introduction To Modern Information Retrieval. McGraw-Hill, New York.

Sankoff, D. and Kruskal, J.B. (1983). Time Warps, String Edits And Macromolecules. The Theory And Practice Of Sequence Comparison. Addison-Wesley, Reading,Ma.

Shepherd, M.J. and Willmot, A.J. (1968). Cluster analysis on the ATLAS computer. Comput. J. 11, 57-62.

Sibson, R. (1973). SLINK: an optimally efficient algorithm for the single-link cluster method. Comput. J. 16, 30-34.

Sneath, P. H. A. and Sokal, R. R. (1973) Numerical Taxonomy. Freeman, San Francisco.

Spath, H. (1980). Cluster Analysis Algorithms. Ellis Horwood, Chichester.

Tou, J.T. and Gonzalez, R.C. (1974). Pattern Recognition Principles. Addison-Wesley, Reading,Ma.

Van Rijsbergen, C.J. (1979). Information Retrieval. Butterworth, London.

Vorhees, E. (1985). The Effectiveness and Efficiency of Agglomerative Hierarchic Clustering in Document Retrieval. Ph.D. thesis, Cornell University.

Weir, A.D. (1970). Program DIVINFRE: A Divisive Classification On Binary Data Including Re-allocation. CSIRO Division of Land Research, Canberra.

Whitney, V.K.M. (1972). Algorithm 422. Minimal spanning tree. Comm. ACM 15, 273-274.

Willett, P. (1980). Document clustering using an inverted file approach. J. Inform. Sci. 2, 223-231.

Willett, P. (1983). Similarity coefficients and weighting functions for automatic document classification: an empirical comparison. Int. Classif. 10, 138-142.

Williams, W.T. and Dale, M.B. (1965). Fundamental problems in numerical taxonomy. Adv. Botan. Res. 2, 35-68.

Williams, W.T. and Lambert, J.M. (1960). Multivariate methods in plant ecology. II. The use of an electronic digital computer for association analysis. J. Ecol. 48, 689-710.

Williams, W.T., Lance, G.N., Dale, M.B. and Clifford, H.T. (1971). Controversy concerning the criteria for taxonomic strategies. Comput. J. 14, 162-165.

Wishart, D. (1969). Mode analysis, a generalization of nearest neighbour which reduces chaining, in Numerical Taxonomy. (Edited by A.J. Cole). Academic Press, London.

Wishart, D. (1978). CLUSTAN 1C User Manual. Program Library Unit, Edinburgh University.

Zahn, C. (1971). Graph-theoretic methods for detecting and describing gestalt clusters. IEEE Trans. Comput. C-20, 68-86.

CHAPTER 3
Similarity in Chemical Information Systems

3.1 THE MEASUREMENT OF STRUCTURAL SIMILARITY

The discussion in Section 2.6 has shown that there are several areas of chemical retrieval that could be improved substantially by the use of nearest neighbour searching. For such techniques to be used, however, some quantitative measure is required to determine the degree of similarity between a query structure and each of the structures in the file being searched.

The concept of inter-molecular structural similarity is not a new one, and there have been a few reports in the literature of methods for the determination of the similarity between pairs of structures. An early definition of similarity is provided by the work of Adamson and Bush (1973), who characterised compounds by lists of the fragments that were present within them. A quantitative measure of inter-molecular similarity could then be obtained by a comparison of two such lists using a distance or similarity coefficient of some sort. A study of simulated property prediction experiments for the pI values of the 20 naturally occurring amino acids demonstrated the basic utility of the approach (Adamson and Bush, 1973), while subsequent work evaluated a wide range of types of similarity coefficient (Adamson and Bush, 1975); further experiments are reported by Bush (1977). This work forms the basis for the studies reported in Section 3.2 below.

Rather than considering structural descriptors which are based upon isolated fragment substructures, Randic and Wilkins (1979) have described a similarity procedure which is based upon the enumeration of all of the inter-atomic paths present within a molecule. This definition of similarity is not restricted to isolated fragment substructures, but takes into account the inter-connections between atoms, and thus the overall topology of a structure. A molecule is characterised by a sequence of integers, each of which corresponds to the numbers of paths of a given length within the structure; this sequence may then be compared with the sequences of other molecules so as to determine the degree of similarity between them. Specifically, let P[I,J] be the number of paths of length J bonds centred upon the I'th atom in a molecule. Then a vector V may be defined, the J'th element of which, V[J], contains the total number of paths of length J bonds in the molecule, i.e.

$$V[J] = \sum P[I,J]$$

where the summation is over all of the N atoms in the structure. The vectors VK and VL for two compounds K and L may be compared to yield a squared distance given by

$$\sum (VK[J]-VL[J])^2$$

where the summation is from J=1 to MAX, the largest path length in the two structures. These distances may be used for the ranking of structures with respect to a specified query molecule. Randic and Wilkins (1979) describe the use of such rankings for the analysis of a set of hypothetical monocyclic monoterpenes, while other applications in QSAR are discussed by Wilkins and Randic (1980) and by Wilkins et al. (1981). A very similar approach has been described by Gabanyi et al. (1982) for the ranking of molecules when compared with lead compounds in drug optimisation programmes. An advantage of this work is that the calculated similarities take account of the atomic types present within a compound, whereas the Wilkins-Randic model is based solely upon topological considerations without regard for the elemental composition of the structures that are being compared.

Related work has been described by Broto et al. (1984) using the mathematical theory of autocorrelation functions. The model assumes that each atom, I, has some associated property or characteristic, F[I]. Then an autocorrelation vector V is defined as

$$V = (V0, V1, ... VJ ... VMAX)$$

where VJ contains the sum of the products F[K]*F[L] for all pairs of atoms, K and L, which are separated by exactly J bonds, 0<=J<=MAX where MAX is the maximum possible inter-atomic separation within that structure. The vector can be used as a characterisation of the structure, and the comparison of two such vectors provides a means for determining the degree of similarity between any two compounds. The individual F[I] values can represent any sort of parameters, e.g., structural, physical or biological, and different autocorrelation vectors may be obtained depending upon the actual parameter type that is chosen. If F[I] is set to the constant value of 1, for all I, then the components of the autocorrelation vector represent the number of paths of lengths 0,1,2.. within the molecule, i.e. the measure described by Wilkins and Randic. Other types of vector discussed by Broto et al. are based upon atomic logP values, van der Waal, atomic radii and inter-atomic distances.

The most extensive development of Randic and Wilkins' work has been described by Carhart et al. (1985) who have described a search system in which structures are characterised by descriptors representing a pair of atoms and the number of bonds separating them. The identification of the descriptors common to a query molecule and each of the structures in the file provides a basis for the ranking of the compounds in order of decreasing similarity with the query. The approach is closely related to that described in Section 3.3.3 below, but differs in the use of the serial file organisation, which does not permit interactive best match searching, and in the types of descriptor that are used. These authors also report the use of the atom pair descriptors in substructural analysis studies. With the exception of this work of Carhart et al., all of these studies have been in the context of SAR, have involved data sets of limited size, and have ignored the possibility of using similarity-based ranking methods as a general search mechanism in computerised structure retrieval systems.

A rather different approach to the identification of chemical similarity is reflected in the work done on the automatic indexing of chemical reactions for storage and retrieval purposes. As noted in Section 1.5, one of the main aims of this work has been to develop efficient algorithms for the identification of those parts of the reactant and product molecules that have not been changed in the course of a reaction (Willett, 1986). The most obvious way to do this involves the identification of the maximal common subgraph, or MCS, for the sets of reactant and product molecules, and several such algorithms have been described (Cone et al., 1977; Varkony et al., 1979). However, such algorithms are extremely demanding of computational resources, and are quite infeasible for the processing of large data bases of reactions. Accordingly, heuristic algorithms have been developed which allow the efficient identification of large common substructures, albeit not necessarily the MCS (Lynch and Willett, 1978; McGregor and Willett, 1980). To date, there has been little attempt to relate such studies of structural overlap to property prediction and structure-based retrieval; however, it may be noted that there have been several reports in the literature discussing the use of 'super-molecules' onto which structures of known activity may be mapped using a superimposition technique (Cammarata and Menon, 1976; Henry and Block, 1979). It may be that MCS-like algorithms could provide an automated means for the identification of such super-molecules, and hence for the degree of structural similarity between groups of chemical compounds, if efficient MCS algorithms could be developed.

The work reported in this and subsequent chapters takes as its basis the quantitative definition of inter-molecular structural similarity used by Adamson and Bush (1975). Specifically, it is assumed that each of the structures under consideration is characterised by a set of fragment substructure descriptors, and that the degree of similarity between a pair of structures can be evaluated in terms of the fragments that are, and that are not, common to both of them.

Many types of similarity measure have been reported in the literature but there seems to have been relatively little discussion as to the merits of different measures, and it is thus of some importance to determine which measures are most appropriate to the determination of inter-molecular structural similarities. This question is addressed in Section 3.2 which reports an extended comparison of similarity measures with their effectiveness being determined by simulated property prediction experiments. The results of these experiments, which are restricted to small sets of compounds, are then used as the basis for the nearest neighbour chemical search systems described in Section 3.3.

3.2 THE EVALUATION OF SIMILARITY MEASURES

3.2.1 Introduction

The determination of inter-molecular similarity necessitates the specification of two functions, these being the weighting scheme which is used to differentiate between different types of fragment, and the similarity coefficient which is used to obtain a numeric quantification of the degree of similarity between a given pair of molecules. As has been noted in Section 2.2, there is an extensive literature associated with both types of function, and the weighting schemes and similarity coefficients used here have been selected from this literature to reflect the very wide range of functions that have been suggested. In all, six weighting schemes and six similarity coefficients are studied, a total of 36 different similarity measures.

The evaluation of the weighting schemes and similarity coefficients involves 16 sets of molecules selected from the structure-activity and structure-property literatures; these data sets are also used for many of the clustering experiments described in Chapter 4. The data sets are as follows.

A. pI values of 20 naturally occurring amino acids (Adamson and Bush, 1973)

B. local anaesthetic activity of 37 structurally diverse compounds (Adamson and Bush, 1975)

C. inhibition of ribonucleotide reductase by 28 substituted benzohydroxamic acids (van't Riet et al., 1980)

D. toxicity to mice of 25 aliphatic and carbocyclic ethers (DiPaolo, 1978a)

E. chromatographic retention indices for 56 aliphatic alcohols, ketones, ethers and esters (Kier and Hall, 1979)

F. inhibition of complement by 105 benzamidines (Hansch and Yoshimoto, 1974)

G. inhibition of monoamine oxidase by 24 hydrazides (Richard and Kier, 1980)

H. chymotrypsin hydrolysis of 72 _N_-acyl esters (Hansch et al., 1977)

I. toxicity to mice of 27 aliphatic hydrocarbons, ethers and ketones (DiPaolo, 1978b)

J. serum binding activity of 79 penicillins (Adamson and Bawden, 1975)

K. antifungal activity of 28 substituted phenyl propyl ethers (Hall and Kier, 1978)

L. inhibition of dihydrofolate reductase by 46 quinazolines (Chen et al., 1979)

M. inhibition of chlorphentermine binding in rat lung by 20 structurally diverse compounds (Minchin and Madsen, 1980)

N. molar refractivity of 65 aliphatic ethers, amines, alcohols and halides (Kier and Hall, 1976)

O. tadpole narcosis activity of 34 structurally diverse compounds (Kier et al., 1975)

P. heats of vaporisation for 129 alkanes, alkenes, alcohols, ketones and benzene and pyridine derivatives (Adamson and Bawden, 1980)

These data sets span a wide range of chemical, physical and biological properties, and include both homogeneous and heterogeneous sets of compounds. It is hoped that the use of such a wide range of types of compound and property will ensure that the results are of some generality, and not conditioned by the characteristics of a particular data set. It may be noted that the number of compounds listed above is marginally less than in the original data set in some cases, specifically sets B, F, M and P. The differences arise from limitations in the connection table software that was used in the experiments.

The compounds are characterised for experimental purposes by sets of substructural descriptors, the primary descriptor type used for the experiments reported in this and subsequent chapters being the augmented atom as described in Chapter 1. For a data set containing N compounds characterised by M fragments, an NxM data matrix, D, is generated where D[I,J] contains the number of occurrences of the J'th fragment in the I'th compound. This matrix may then be used to generate a similar matrix, W, where W[I,J] is the weight of the J'th fragment in the I'th compound using one of six weighting schemes. Each such data matrix forms the basis for the generation of a symmetric NxN similarity matrix, S, the elements of which, S[K,L], correspond to the similarity between molecules K and L using one of six similarity coefficients. The weighting schemes and similarity coefficients tested are described below in Section 3.2.2.

The effectiveness of each similarity matrix, S, in characterising the inter-molecular relationships present in a data set may be assessed by simulated property prediction experiments using a 'leave-one-out' approach. For each molecule K, the set of N-1 elements

S[K,1], S[K,2]..S[K,L]..S[K,N] K <> L

in S is scanned to identify the molecule, A say, that is most similar to K on the basis of the selected similarity measure. The predicted property value for compound K, P[K], is then set equal to the observed property value for compound A, O[A], and the overall measure of agreement between the N observed and predicted property values determined by means of the product moment correlation coefficient. Since 36 different similarity measures are tested, a total of 36 correlation coefficients is obtained from D for each one of the 16 data sets studied.

An entirely comparable set of 36x16 experiments uses as the evaluation criterion the mean of the differences between the observed and the predicted values, i.e.

$$\frac{\sum |P[K]-O[K])|}{N}$$

3.2.2 Similarity measures

Several types of statistical information can be extracted readily from machine-readable representations of chemical structures to form the basis for a fragment weighting scheme.

The simplest weighting scheme involves a consideration of the number of occurrences of each particular fragment type within a compound, with frequently occurring fragments being given a greater weight than those that occur less frequently: thus a pair of molecules that had several occurrences of a given fragment in common would be considered as more similar to each other than if they had only a single occurrence in common. The second type of weighting takes into account the sizes of the molecules that are being compared, where by size is meant the number of fragments assigned to a compound. Thus a fragment in a small molecule would be assigned a greater weight than the same fragment in a larger compound, and fragments common to a pair of small molecules would be regarded as a better indicator of structural similarity than the sharing of a common fragment by two large structures. The final type of weighting scheme involves the inverse frequency of occurrence of a fragment in the data set, with infrequently occurring fragments being assigned greater weights than those that occur throughout the data set: thus a pair of compounds that both possessed some rare substructural feature would be regarded as being much more strongly related to each other than if they both possessed a feature such as a benzenoid ring.

The weighting schemes tested here reflect these three types of considerations. Each of these schemes seems to be intuitively quite reasonable, and there seems to be no obvious a priori reason why any one of them should be preferred to the others.

A fourth type of weighting scheme, which has been used in the substructural analysis studies described in Section 1.4, takes into account the differential occurrences of substructures in the active and inactive compounds within a data set. Such occurrence information, however, pre-supposes that several structures have already been analysed in some detail to provide the requisite activity data, whereas the main thrust of the work reported here is aimed towards the identification and retrieval of compounds when such information is not available.

The first 3 weighting schemes, W1 to W3, consider only the presence or absence of a particular fragment descriptor in a compound without regard to its frequency of occurrence, so that all of the D[I,J] values are either 0 or 1. The three schemes are as follows:

$$W1 : W[I,J] = D[I,J]$$

which corresponds to the use of completely unweighted structural descriptions;

$$W2 : W[I,J] = \frac{1}{\sum D[I,J]}$$

where the summation is over all J, this corresponding to a normalisation by the number of fragments present in a molecule; and

$$W3 : W[I,J] = \log_e \left[\frac{N}{\sum D[I,J]} \right]$$

where the summation is over all I, this corresponding to the inverse weighting scheme above; the use of a logarithmic relationship is suggested by the work of Harrison (1968).

The second set of weighting schemes, W4 to W6, is identical to the set W1 to W3, except that the original frequencies are used, without conversion into a 0-1 scale.

The discussion of similarity coefficients in Section 2.2 has noted that there are four main types of coefficient, these being association measures, distance measures, correlation coefficients, and probabilistic coefficients. The experiments here use examples of the first three types of coefficient: probabilistic coefficients are not considered in view of the results of Adamson and Bush (1975) who suggested that such coefficients are of little value for this application area. The similarity coefficients used are as follows: their mathematical formulations have been given in Section 2.2.2 and will thus not be repeated here:

S1 : cosine coefficient

S2 : Tanimoto coefficient

S3 : Euclidean squared distance

S4 : correlation coefficient

S5 : Canberra distance

S6 : average distance

As Sparck Jones (1973) notes, the weighting scheme and the similarity coefficient used for the calculation of a similarity measure may be related. Thus the effect of the denominator in a coefficient like S1, the cosine coefficient, is clearly akin to the effect of a weighting scheme like W2 or W4 since both of them take account of the number of fragments in a compound, and it may thus be argued that certain of the combinations of weighting function and similarity coefficient are redundant. However, all of the 36 possible combinations of similarity coefficient and weighing function are tested here, in the interest of completeness.

3.2.3 Empirical comparisons of similarity measures

The calculated correlation coefficients obtained using each combination of similarity coefficient and weighting function are listed in Table 3.1 on the following pages. Each 6x6 sub-table contains the results for the data set noted in the left-hand margin: the sub-tables are arranged so that the entries in the I'th row and the J'th column contain the product moment correlation coefficient obtained using SI as the similarity coefficient and WJ as the weighting scheme. Thus the coefficient obtained from the use of S4 and W5 on data set C, the 28 benzohydroxamic acids, is 0.42. Coefficients that are not significant at the 0.05 level of statistical significance are replaced by an asterisk.

	0.44	0.44	0.44	0.65	0.78	0.66
	0.44	0.44	0.44	0.68	0.64	0.64
A	*	*	*	0.81	0.67	0.81
	*	*	*	0.69	0.65	0.65
	*	*	*	0.64	*	0.64
	*	*	*	0.81	0.66	0.81
	0.55	0.55	0.52	0.60	0.60	0.61
	0.67	0.71	0.51	0.81	0.76	0.80
B	0.56	0.53	0.48	0.75	0.60	0.76
	0.38	0.60	0.52	0.61	0.73	0.62
	0.56	0.53	0.56	0.52	0.58	0.52
	0.56	0.67	0.56	0.79	0.73	0.73
	0.51	0.51	0.52	0.53	0.53	0.47
	0.51	0.51	0.52	0.49	0.49	0.47
C	*	*	*	*	*	*
	0.51	0.44	0.52	0.53	0.42	0.48
	*	*	*	*	*	*
	*	*	*	*	*	*
	*	*	*	0.65	0.65	0.51
	*	*	*	0.74	0.61	0.65
D	*	*	*	0.49	0.57	0.38
	*	*	*	0.65	0.69	0.51
	*	*	*	*	0.42	*
	*	*	*	0.49	0.61	*

	0.63	0.63	0.62	0.70	0.70	0.69
	0.63	0.63	0.62	0.68	0.70	0.69
E	0.27	0.30	*	0.62	0.63	0.60
	0.60	0.63	0.58	0.66	0.68	0.66
	0.27	*	0.27	0.56	0.65	0.56
	0.27	0.29	*	0.63	0.67	0.60
	0.77	0.77	0.70	0.74	0.74	0.68
	0.77	0.77	0.71	0.83	0.72	0.80
F	0.82	0.75	0.73	0.86	0.71	0.79
	0.78	0.74	0.70	0.75	0.81	0.68
	0.82	0.83	0.82	0.78	0.80	0.78
	0.82	0.81	0.77	0.84	0.76	0.83
	0.79	0.79	0.77	0.68	0.68	0.70
	0.79	0.79	0.77	0.67	0.64	0.70
G	0.77	0.77	0.67	0.75	0.62	0.59
	0.80	0.79	0.77	0.68	0.64	0.70
	0.77	0.72	0.77	0.77	0.73	0.77
	0.77	0.74	0.50	0.79	0.66	0.66
	0.78	0.78	0.81	0.89	0.89	0.87
	0.78	0.78	0.81	0.88	0.89	0.89
H	0.77	0.83	0.75	0.91	0.91	0.90
	0.78	0.86	0.81	0.90	0.89	0.87
	0.77	0.77	0.77	0.81	0.88	0.81
	0.77	0.84	0.75	0.88	0.93	0.88
	*	*	*	0.78	0.78	0.78
	*	*	*	0.76	0.78	0.78
I	*	0.39	*	0.78	0.78	0.79
	*	*	*	0.78	0.82	0.78
	*	*	*	0.82	0.78	0.82
	*	0.39	*	0.78	0.78	0.79

	0.60	0.60	0.56	0.81	0.81	0.75
	0.60	0.60	0.55	0.79	0.77	0.78
J	0.49	0.45	0.38	0.74	0.72	0.71
	0.63	0.64	0.56	0.80	0.71	0.75
	0.49	0.45	0.49	0.62	0.73	0.62
	0.49	0.56	0.38	0.74	0.70	0.68
	0.42	0.42	0.42	0.63	0.63	0.65
	0.42	0.42	0.42	0.63	0.63	0.65
K	*	*	*	*	*	*
	0.42	*	0.42	0.63	0.54	0.65
	*	*	*	*	*	*
	*	*	*	*	*	*
	0.83	0.83	0.84	0.82	0.82	0.82
	0.83	0.83	0.84	0.81	0.81	0.82
L	0.77	0.81	0.77	0.72	0.70	0.72
	0.83	0.84	0.84	0.82	0.81	0.82
	0.77	0.80	0.77	0.82	0.82	0.82
	0.77	0.75	0.73	0.65	0.78	0.87
	0.55	0.55	*	0.44	0.44	*
	0.55	0.55	0.37	0.51	0.45	*
M	0.58	*	*	0.52	*	*
	0.55	*	*	0.46	*	*
	0.58	0.57	0.58	0.49	0.51	0.49
	0.58	0.70	*	0.66	0.46	0.53
	0.28	0.28	0.29	0.54	0.54	0.62
	0.28	0.28	0.29	0.65	0.56	0.61
N	0.38	0.38	0.33	0.63	0.40	0.56
	0.28	0.27	0.29	0.54	0.65	0.62
	0.38	0.28	0.38	0.50	0.55	0.50
	0.38	0.38	0.33	0.56	0.36	0.56

	0.45	0.45	0.46	0.71	0.71	0.72
	0.45	0.45	0.46	0.72	0.85	0.90
O	*	0.40	*	0.69	0.84	0.64
	0.46	0.45	0.46	0.71	0.91	0.72
	*	*	*	*	0.49	*
	*	0.37	*	0.35	0.79	0.50
	0.82	0.82	0.81	0.87	0.87	0.92
	0.82	0.82	0.81	0.80	0.92	0.91
P	0.71	0.75	0.72	0.81	0.81	0.84
	0.81	0.82	0.81	0.87	0.90	0.92
	0.71	0.69	0.71	0.80	0.82	0.80
	0.71	0.68	0.68	0.80	0.80	0.86

Table 3.1

Product moment correlation coefficients between observed and predicted property values for the 16 data sets. Starred values are not significant at the 0.05 level of significance.

An inspection of the coefficients in Table 3.1 shows that, in general, fair correlations are obtained between the observed and predicted property values. The extent of this correlation naturally varies from one data set to another, the chymotrypsin hydrolysis, quinazoline and heats of vaporisation data sets, for example, giving high correlations with all weighting schemes and similarity coefficients. In some cases, the correlations are generally low, but even here there are usually a few similarity measures that give more acceptable results; examples of this behaviour are provided by the amino acid, ether toxicity and phenyl propyl ether data sets.

Of more importance in the context of a comparative study such as this are the differences between similarity measures within a data set, and across the entire range of data sets. The primary subjective impression from an inspection of the results in Table 3.1 is that the weighting schemes W4 to W6 tend to give a noticeably higher relationship between structure and property than do the functions W1 to W3, which ignore the actual fragment frequencies of occurrence in favour of a simple binary molecular characterisation. Examples of this behaviour include the amino acid, gas chromatography, penicillin, refraction and tadpole narcosis data sets inter alia; conversely, other data sets, such as the benzohydroxamic acids and the hydrazides, exhibit little or no difference between the various schemes.

Rather less difference is evident between the six similarity coefficients tested, although the distance measures sometimes perform less effectively than the the other three coefficients. Examples include the amino acid, aryl acid, gas liquid chromatography, phenyl propyl ether and tadpole narcosis data sets; the guinea pig complement figures are an exception to this general rule.

A statistical analysis of the results in these tables, using the Kendall coefficient of concordance (Siegal, 1956), is described by Willett and Winterman (1986) to test the hypothesis that there are no significant differences between the effectiveness of the six weighting schemes when the similarity coefficient is fixed. The analysis rejected this hypothesis and confirmed the subjective impressions engendered by an inspection of Table 3.1 since it was found that small, but significant, correlations do indeed exist between the rankings, and that the schemes W4 to W6 yield better correlations than do the schemes W1 to W3 based upon binary data matrices. An entirely comparable series of tests was carried out to test the hypothesis that there are no significant differences between the effectiveness of the six similarity coefficients when the weighting scheme is fixed. These results confirm a significant, but small, relationship between the 16 sets of rankings, and poorer correlations, when the three distance measures, S3, S5 and S6, are used as the similarity coefficient.

The results from the mean difference tests are shown in Table 3.2.

	0.45	0.45	0.42	0.95	0.68	0.95
	0.45	0.45	0.42	0.64	0.94	0.94
A	0.51	0.40	0.88	0.29	0.71	0.28
	0.39	0.23	0.53	0.85	0.95	0.95
	0.51	0.09	0.51	0.24	0.97	0.24
	0.51	0.41	0.88	0.26	0.53	0.26
	0.32	0.32	0.18	-0.03	-0.03	-0.16
	0.67	0.57	0.42	-0.32	-0.06	-0.16
B	-0.76	0.84	-1.15	-0.50	0.43	-0.56
	-0.01	1.27	0.13	-0.21	0.79	-0.20
	-0.76	-0.77	-0.76	-0.42	-0.54	-0.42
	-0.76	0.19	-0.99	-0.32	0.24	-0.48
	-0.53	-0.53	-0.52	0.08	0.08	0.05
	-0.53	-0.53	-0.52	0.06	0.06	0.04
C	-0.11	-0.12	0.08	-0.18	-0.09	-0.10
	-0.53	-0.53	-0.52	0.07	0.02	0.05
	-0.11	-0.08	-0.11	-0.09	-0.09	-0.09
	-0.11	-0.07	0.08	-0.15	-0.13	-0.08

	-0.09	-0.09	-0.07	-0.04	-0.04	-0.00
	-0.09	-0.09	-0.07	-0.09	-0.05	0.03
D	-0.03	-0.08	0.20	0.02	-0.06	0.06
	-0.09	-0.18	-0.07	-0.05	-0.14	-0.01
	-0.03	0.13	-0.03	0.00	0.02	0.00
	-0.03	-0.13	0.20	0.02	-0.03	0.06
	-12.29	-12.87	-12.59	-5.29	-5.40	-4.78
	-12.29	-12.87	-12.59	-6.19	-5.40	-4.93
E	-0.73	-1.11	3.37	-3.98	-0.28	-2.26
	-10.79	-12.42	-11.46	-4.04	-6.20	-3.78
	-0.73	4.98	-0.73	-4.39	-2.24	-4.39
	-0.73	-0.34	3.37	-2.53	-3.56	-2.26
	-0.17	-0.17	-0.16	-0.14	-0.14	-0.17
	-0.17	-0.17	-0.17	-0.10	-0.15	-0.12
F	-0.27	-0.27	-0.12	-0.19	-0.21	-0.12
	-0.13	-0.23	-0.16	-0.14	-0.16	-0.17
	-0.27	-0.20	-0.27	-0.09	-0.10	-0.09
	-0.27	-0.29	-0.09	-0.17	-0.19	-0.12
	-0.02	-0.02	0.01	0.05	0.05	0.01
	-0.02	-0.02	0.01	0.06	-0.02	0.01
G	-0.01	0.03	0.05	0.11	0.16	0.09
	0.03	-0.02	0.01	0.01	-0.02	0.01
	-0.01	-0.05	-0.01	0.19	0.18	0.19
	-0.01	0.05	0.05	0.09	0.19	0.08
	-0.14	-0.14	-0.10	-0.09	-0.09	-0.09
	-0.14	-0.14	-0.09	-0.03	-0.08	-0.11
H	-0.18	-0.18	0.30	0.01	0.01	0.09
	-0.09	-0.22	-0.10	-0.09	-0.14	-0.09
	-0.18	0.03	-0.18	0.06	0.04	0.06
	-0.18	-0.27	0.30	0.02	-0.04	0.06

	-0.61	-0.61	-0.61	-0.03	-0.03	-0.03
	-0.61	-0.61	-0.61	-0.08	-0.03	-0.03
I	-0.58	-0.49	-0.45	-0.11	-0.03	0.03
	-0.61	-0.61	-0.61	-0.03	-0.03	-0.03
	-0.58	-0.58	-0.58	-0.06	-0.03	-0.06
	-0.58	-0.49	-0.45	-0.03	-0.03	0.03
	-0.09	-0.09	-0.08	0.06	0.06	0.04
	-0.09	-0.09	-0.08	0.07	0.07	0.04
J	0.00	-0.03	0.01	0.05	0.02	0.14
	-0.11	-0.07	-0.08	0.06	0.00	0.04
	0.00	0.08	0.00	0.03	0.01	0.03
	0.00	0.03	0.01	0.07	0.02	0.11
	-0.25	-0.25	-0.25	0.09	0.06	0.04
	-0.25	-0.25	-0.25	0.09	0.10	0.04
K	-0.42	-0.40	-0.42	-0.16	0.03	0.09
	-0.25	-0.27	-0.25	0.06	0.05	0.04
	-0.42	-0.19	-0.42	-0.01	0.00	0.01
	-0.42	-0.19	-0.42	0.16	-0.13	0.09
	-0.01	-0.01	-0.10	0.00	0.00	-0.03
	-0.01	-0.01	-0.10	0.00	0.00	-0.03
L	0.20	0.16	0.11	-0.31	-0.32	-0.32
	-0.01	-0.10	-0.10	0.00	-0.03	-0.01
	0.20	0.10	0.20	-0.08	-0.13	-0.08
	0.20	0.23	0.14	-0.24	-0.17	-0.18
	5.24	5.24	2.21	1.65	1.65	-0.45
	5.24	5.24	2.55	3.65	1.59	-1.06
M	-0.80	2.92	-3.23	6.26	-1.48	-0.16
	3.33	1.72	2.21	0.61	0.21	-0.45
	-0.80	-1.13	-0.80	-4.75	-4.40	-4.75
	-0.80	-2.13	-3.23	8.48	2.76	2.58

	-4.62	-4.62	-4.27	-0.90	-0.90	-0.90
	-4.62	-4.62	-4.27	-0.85	-0.82	-0.84
N	-3.78	-3.76	-0.79	0.27	-0.77	0.84
	-4.41	-4.25	-4.27	-0.76	-2.07	-0.90
	-3.78	-0.53	-3.78	-0.93	-0.14	-0.93
	-3.78	-4.12	-0.86	0.55	-0.77	0.84
	-0.46	-0.46	-0.43	0.02	0.02	0.03
	-0.46	-0.46	-0.43	0.03	-0.02	-0.03
O	0.45	0.30	0.48	0.13	0.04	0.14
	-0.44	-0.60	-0.43	0.02	-0.10	0.03
	0.45	0.45	0.45	0.12	0.25	0.12
	0.45	0.17	0.47	0.35	0.12	0.29
	-0.62	-0.62	-0.58	-0.03	-0.03	-0.08
	-0.62	-0.62	-0.58	0.12	-0.10	-0.09
P	-0.38	-0.44	0.15	0.00	-0.03	0.37
	-0.60	-0.56	-0.58	-0.03	-0.10	-0.08
	-0.38	0.12	-0.38	0.06	0.20	0.06
	-0.38	-0.22	0.23	0.07	0.05	0.33

Table 3.2

Mean differences between observed and predicted property values for the 16 data sets.

The results in this table again illustrate the binary schemes, W1 to W3, exhibiting poorer correlations than the schemes W4 to W6 based upon weighted molecular characterisations. When the weighting scheme is fixed, in the significance test mentioned previously, only two of the Kendall values are significant, these being obtained with the inverse logarithmic schemes, W3 and W6; of the coefficients, the best results using this evaluation criterion are given by the Euclidean and average distances. Further details of the experiments are given by Willett and Winterman (1986) who also report the results that were obtained when a rather larger type of bond-centred substructure was used to characterise the compounds in the data sets; this fragment type gave results that were slightly inferior to those given by the augmented atom descriptors.

Several conclusions may be drawn from these experiments. The first, and perhaps most important, general result is that the use of simple, fragment-based, similarity measures enables the identification of structures that are closely related in activity to a specified query molecule. Thus the identification of a single interesting compound may be sufficient to lead to several others in a machine-readable file that might be expected to exhibit comparable activities. Secondly, there would seem to be only slight differences in performance between the similarity coefficients used, with the cosine, Tanimoto and correlation coefficients giving rather better results than the three distance measures when the product moment correlation coefficient is used to evaluate the experiments, but with the distance measures performing well in the mean difference analyses. Accordingly, there would seem to be little basis for preferring one coefficient to another, and thus the selection of a coefficient for some particular application might most usefully be carried out on grounds such as the ease of computation. Finally, it would seem that the most generally useful form of molecular representation based on substructural fragments is a list of the fragments present, together with their frequencies of occurrence within that structure. The use of weighting schemes based upon the frequencies of occurrence of fragment within a data set, or the numbers of fragments within a molecule, do not seem to be of noticeable general utility. These findings form the basis for the searching experiments described in the next section.

3.3 BEST MATCH SEARCHING IN CHEMICAL STRUCTURE FILES

3.3.1 Introduction

The study of similarity measures described in the previous section forms a necessary pre-requisite for the use of such measures in best match searches of large files of chemical structures. As noted in Chapter 1, current chemical structure search systems have two main modes of operation, these being substructure and structure, or registration, searching: similarity-based techniques can be used to enhance both types of retrieval mechanism.

When a chemist carries out a substructure search, there is usually some group of compounds that the search is intended to retrieve; however, a general or fuzzy query, or the lack of sufficiently discriminating fragment screens, may result in the retrieval of many molecules in addition to those that were expected, and it is not generally possible to predict the number of hits that will be obtained in any given search *a priori.* In patent searches, for example, complete recall is essential and the identification of several hundreds of molecules containing the query substructure may well be of importance; more generally, however, the chemist will be interested in a much smaller number of molecules. As has been noted in Section 2.6, there are two rather different, and complementary, approaches which could be used to overcome these problems. The first of these, and the one described in this chapter, involves the use of a *target structure*, this being a typical representative of the class of molecules that the search is intended to retrieve. The compounds retrieved in the substructure search could then be matched individually against the target structure to obtain a ranked list, with the most similar molecules being displayed first to the chemist. The second approach involves the use of a clustering method to group the output into a number of clusters so as to obtain an over-view of the main structural classes present in the output: this is evaluated in Section 4.4.

In the case of a registration search the chemist performs a search for identity to determine whether the query compound is present in the file. A chemist may additionally be interested in compounds that are closely related to the query molecule in structural terms, but may not wish, or may not be able, to specify precisely the nature of these relationships. An example of such a requirement might be at an early stage in a drug development program where a potential lead compound is available, but where the substructural requirements for activity have not been located precisely. Such a query could be handled in current systems only by performing a whole series of substructure searches, each of which would correspond to a possible topological pharmacophore. A best match retrieval system, conversely, would allow the molecule of interest to be matched against the entire file to obtain either some number of the most similar structures, or all structures whose similarities with the query compound were greater than some threshold value. Such a system would facilitate browsing through a data base since the discovery of some interesting structure, whether because of its structural characteristics, biological activity or whatever, could then be used as the basis for a subsequent search.

This section describes the implementation of such nearest neighbour search facilities in a conventional chemical structure retrieval system, specifically that of the Research Information Services Department of Pfizer Central Research, Sandwich, UK. All of the examples that are given have been obtained from searches of a file of circa 8,000 compounds, this being a sub-set of the main structure file which contains about 240,000 molecules. In this file, the structures are stored for search as a bit map B in which the bit B[I,J] is set to be true if the J'th screen is present in the I'th compound. The screens are atom-centred or bond-centred fragment substructures which have been selected for inclusion in the screen set on the basis of their frequencies of occurrence in the file of compounds; in all, 1315 screens are available for the characterisation of the molecules in the file. The file is organised to facilitate rapid access to the columns of the bit matrix, i.e. to the lists of compounds that contain particular screens. Substructure searching is accomplished efficiently by the intersection of the lists corresponding to the screens that have been assigned to the query substructure. Full details of the work are given by Willett et al. (1986).

3.3.2 Ranking of substructure search output

From the experiments reported in Section 3.2 it would seem that the most useful form of molecular characterisation for similarity searching is a list of the augmented atoms present within a compound, together with their frequencies of occurrence. However, for reasons of search efficiency, the first-level substructure search is carried out using bit string screening and atom-by-atom routines in the conventional manner, this search resulting in the identification of those molecules containing the query substructure. The second-level search is then carried out to discriminate between this sub-set of the data base to identify those compounds most similar to the target structure. Accordingly, once a set of structurally related molecules has been retrieved by the substructure search, the connection tables for these compounds are used to generate sets of augmented atom fragments to act as a basis for the ranking procedure. A corresponding set of fragments is also generated for the target compound that is to act as the basis for the ranking. The sets of fragments are used to rank the molecules on the basis of their similarity with the target, and the most similar compounds may then be displayed on a graphics terminal to the chemist in order of decreasing similarity. The Tanimoto coefficient was used for the generation of the rankings since early experiments showed that it gave rise to rankings that were more intuitively acceptable than those arising from the use of the cosine coefficient; an explanation of this behaviour is given by Willett et al. (1986).

An example of the use of the system is shown on the next page in Figure 3.1 where the target structure I is used to rank a set of mono-substituted epoxide rings; the four most similar structures are as shown, and it will be seen that the ranking procedure has been successful in identifying structures that would appear intuitively to be very similar in character to the target compound.

Experiments with a range of fragment types suggest that the best match search is dependent upon the types of fragment used for the characterisation of the compounds in the substructure search output. The rankings obtained are generally more satisfactory the larger the size of the fragments that are used, with the best results being obtained consistently with the augmented atom fragments. This finding is in line with the many previous studies of

I

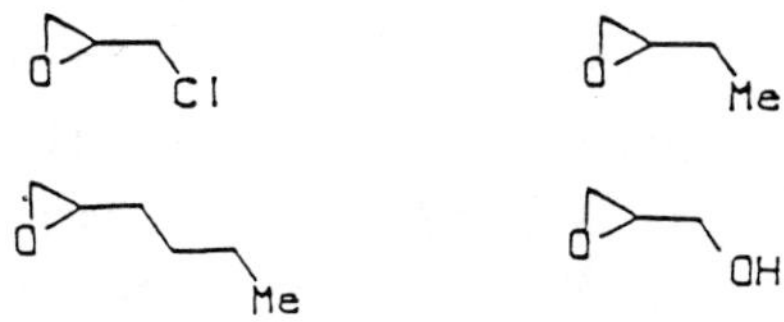

Figure 3.1

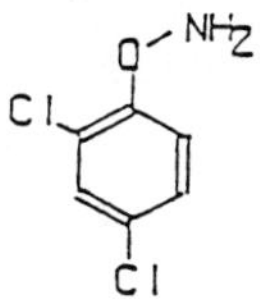

II

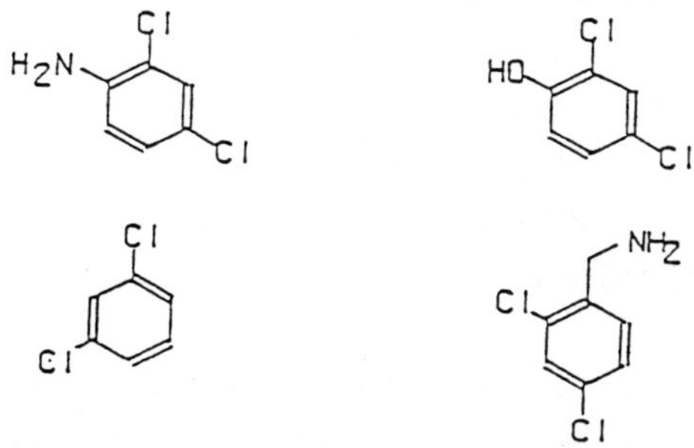

Figure 3.2 (a)

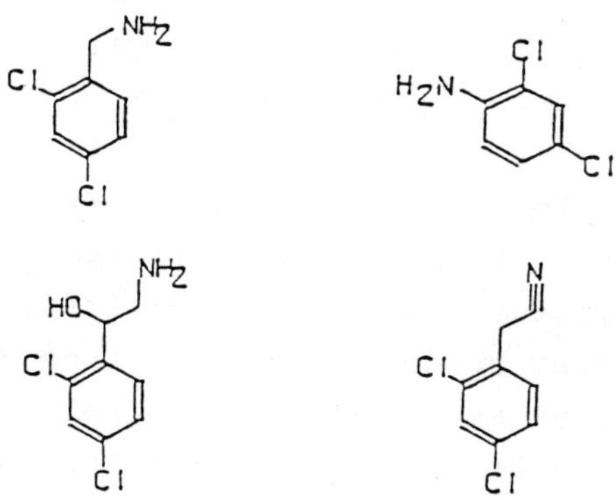

Figure 3.2 (b)

substructural analysis methods as described in the review by Bawden (1983). However, while large substructures such as the augmented atom help to identify the detailed bonding pattern within a molecule, smaller and less specific fragments are useful in characterising the molecular size and composition. This would suggest that superior results would be obtained by using a combination of several types of atom-centred fragment. Thus, the use of the target structure II on a set of 1,3-dichlorobenzenes identifies the four nearest neighbour compounds in Figure 3.2 (a) when augmented atoms are used. However, the use of the augmented atoms in conjunction with the simple, coordinated and bonded atom fragments described in Section 1.3 gives the compounds shown in Figure 3.2 (b), a result that is felt to be more intuitively acceptable.

It is often helpful to be able to differentiate between cyclic and acyclic bonds during the generation of the fragments. Thus, the use of the target structure III on a set of amine derivatives yields the compounds of Figure 3.3 (a) when the combined set of four atom-centred fragment types is used and the bond types are not differentiated. When ring and acyclic bonds are differentiated, however, the nitrogen ring structure is replaced by the more acceptable acyclic structure in Figure 3.3 (b). In other cases, however, cyclic and acyclic structures are strongly differentiated in the rankings even without the use of this refinement. Thus, the use of the target structure IV in Figure 3.4 on a set of structurally diverse organo-phosphorus compounds gives a ranking consisting of three discrete sub-sets, these consisting of tri-phenyl, mono-phenyl, and acyclic molecules. For some purposes, of course, it may be acceptable or helpful to consider ring and chain bonds

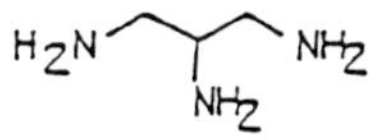

III

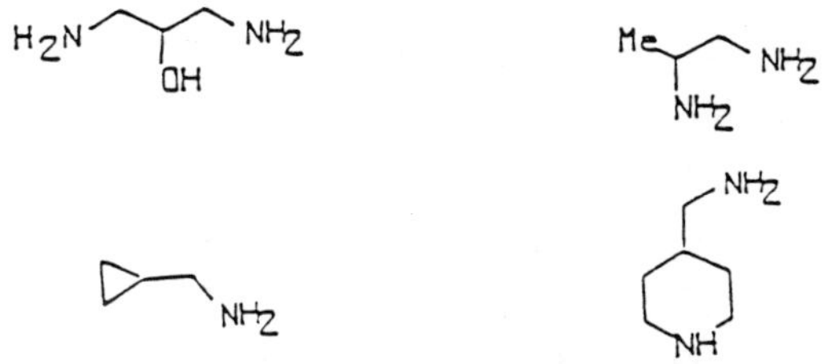

Figure 3.3 (a)

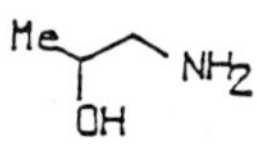

Figure 3.3 (b)

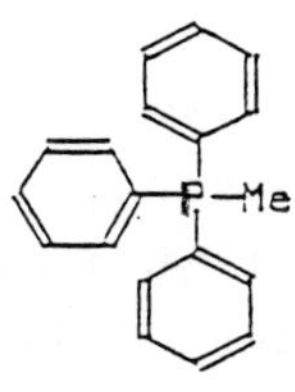

IV

Figure 3.4

as synonymous, and it would accordingly seem that differentiation between ring and chain bonds should be made available as a run-time option for specification as a particular query may dictate.

Many substructure searches are sufficiently specific in nature to result in the retrieval of only a few molecules that satisfy the constraints of the query; when this is not so, however, it is clear that ranking procedures of the sort described here can provide a simple means of focussing upon the structures that are of greatest interest. The experiments have been carried out solely within the context of internal chemical files, but, with the increasing availability of downloading facilities (Mortensen, 1984; Gray, 1985), there is no reason in principle why the ranking methods could not be applied to the output from substructure searches of public chemical data bases such as CAS ONLINE.

3.3.3 Browsing in structural data bases

Whereas the procedures described in the previous section provide a means of modifying the output from a conventional substructure search system, this section outlines a nearest neighbour modification of registration, or structure, search which provides a browsing mechanism for data bases of chemical structures. The major problem associated with the provision of such facilities is that of ensuring a sufficiently rapid response to permit interactive searching. The approach described in the previous section for the ranking of substructure search output is efficient in operation since the fragment generation and similarity matching procedures are applied to the small numbers of molecules retrieved by the substructure searches. The procedure is relatively slow in operation and such methods could accordingly be applied to a whole structure file only if an interactive response was not required; an example of a batch-oriented similarity searching system is described by Carhart et al. (1985). However, there has recently been considerable interest in novel nearest neighbour search algorithms that permit the efficient identification of the compound or compounds most similar to some specified query structure. This work is described in detail in Chapter 5 below and one of these algorithms, algorithm F there, was selected for the implementation of the system described here.

A chemist wishing to use the system inputs the structure of interest at a graphics terminal using a light pen to select and sketch the structure. The fragment screens corresponding to this structure are generated automatically and then used as the input to the nearest neighbour search procedure. The algorithm assumes the presence of an inverted file to the compounds that are to be searched and is based upon the addition of the inverted file lists corresponding to the fragments in the query. The addition results in a vector, the elements of which contain the number of fragments in common between the query structure and each of the structures in the file. Given this information, it is simple to evaluate an inter-molecular similarity coefficient, specifically the Tanimoto coefficient, and then to rank the compounds in order of decreasing similarity with the query structure. The most similar structures are then displayed to the chemist on the graphics terminal. The procedure is sufficiently fast in operation to allow interactive best match searching of large files of structures, with a response time comparable to that for conventional substructure searching; a full description of this algorithm is given in Section 5.3.3.

Two points need to be made at this point about the approach. Firstly, since it is to be applied to an inverted file that contains only binary fragment data, i.e the bit map B, the resulting inter-molecular similarities are likely to be a less accurate reflection of the similarities between file compounds and the chosen target structure than if full fragment occurrence data is available. As noted in Section 5.5 the algorithm can be used with such information but its inclusion in the the bit map would result in considerable storage overheads. Secondly, and unlike the procedure described in the previous section, the methods can be used only where ready access is available to the inverted file posting lists: while this will generally be true of internal structure files, the procedure would not seem to be directly applicable to the searching of public chemical data bases.

The nearest neighbour search is carried out in two stages as with the substructure search procedure described in Section 3.3.2; however, both of the stages in this case involve the ranking of compounds.

The first stage involves the processing of the inverted file lists corresponding to the screens assigned to the target query molecule, which we shall call query screens. These lists are added together as if they were vectors so as to identify the numbers of screens in common between the query molecule and each of the structures in the file. These numbers form the basis for a display which shows the numbers of molecules that contain a certain pre-specified percentage of the query screens, thus allowing the chemist who is carrying out the search to obtain a feeling for the distribution of similarities present. To illustrate this, the query structure IV which has been used previously to exemplify the workings of the ranked substructure search, is assigned a total of 34 different screens and yields an output of the form

P :	100	90	85	80	75	50	25
N(P) :	12	35	45	47	451	3818	4648

where N(P) is the number of structures in the file that contain at least P% of the query screens. This display, which is typical of those obtained for a wide range of types of query molecule, reveals a highly skewed distribution of similarities with relatively few molecules having a high degree of similarity with the query, but with the numbers increasing very rapidly as P is decreased so that by the time P=25 is reached, N(P) corresponds to over half of the file. For this particular example, it seems that the great majority of the structures of potential interest to the chemist are likely to be contained within the 451-member sub-set of the file that contains at least 75% of the query screens. This sub-set may then be used for ranking purposes in the second-stage search, which involves the calculation of the degree of similarity between the target compound and each of the compounds in the sub-set. The Tanimoto coefficient is again used as the basis for the rankings. Since only binary molecular characterisations are used here, i.e. the fragment bit strings, the coefficient may be calculated as C/(SIZE+QSIZE-C) for a structure containing SIZE screens, C of which are in common with a query which has been assigned a total of QSIZE screens. When the ranking has been generated for these 451 compounds, the molecules shown in Figure 3.5 are identified as the four nearest neighbours for the target structure.

A further example is provided by the target compound V in Figure 3.6 which gives an output

N(P) :	0	0	4	4	26	1602	4878

Sorting the 1602 compounds in the P=50 sub-set identifies

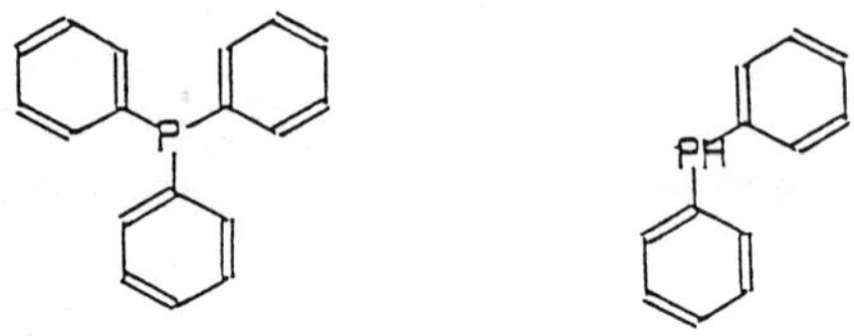

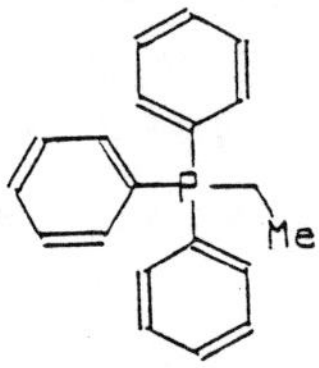

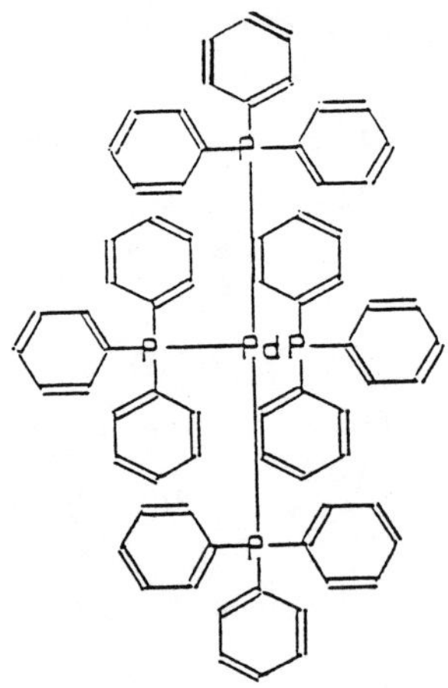

Figure 3.5

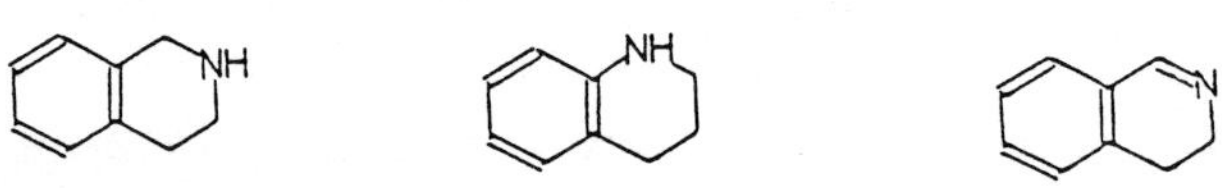

V

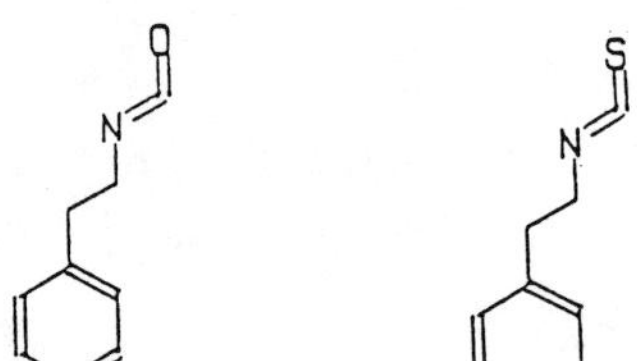

Figure 3.6

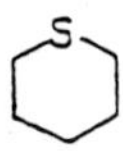

VI

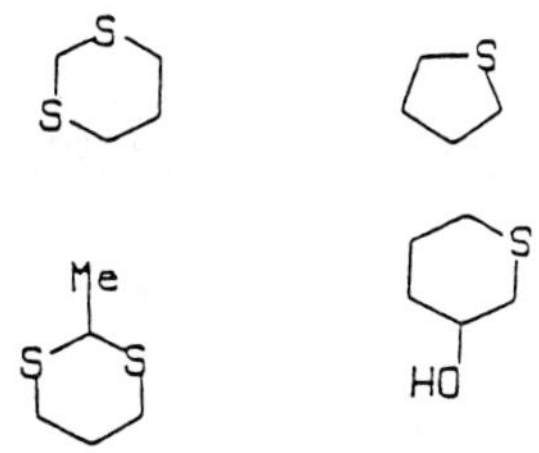

Figure 3.7

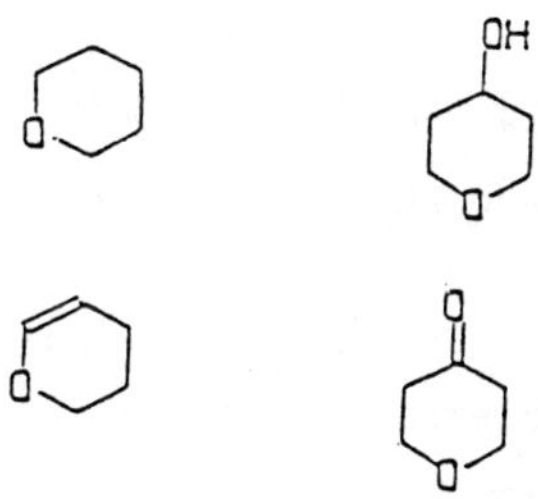

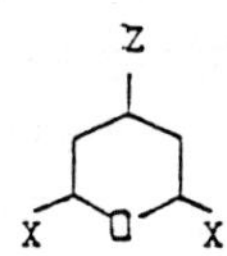

VII

Figure 3.8

the four molecules in Figure 3.6 as the nearest neighbours. Another example of a structure for which there are very few closely-related molecules is provided by VI which gives

N(P) :	1	8	10	13	29	180	5188

The second-stage search with P=50 results in the four heterogeneous compounds shown in Figure 3.7. As a final example, consider the substructure VII, where Z is any atom and X is any atom other than hydrogen; the output is

N(P) :	42	127	177	232	441	2165	6610

with the P=75 sub-set resulting in the identification of the four nearest neighbours shown in Figure 3.8.

As well as the procedure described here, Willett et al. (1986) describe an alternative nearest neighbour search algorithm involving a rather different second-level search. Specifically, once the P-N(P) plot has been obtained in response to the first-level search, a note is made of each distinct value of C, the number of screens common to the target compound and a structure in the file, and then the Tanimoto coefficient is calculated for all compounds having each particular value of C; this corresponds to a ranking of the structures containing C of the query screens in order of increasing SIZE. The overall ranking is then obtained by concatenating the rankings resulting from C=QSIZE, QSIZE-1, QSIZE-2 etc., until all of the compounds in the chosen sub-set, e.g. P=75, have been ranked. For some target molecules, this alternative retrieval mechanism results in a rather different ranking from those presented above, thus providing alternative sets of compounds for consideration by the chemist who is running the search. Additional flexibility may be obtained by varying the threshold parameter P; it is, of course, possible to set P to a very low value, such as P=25 or even P=0, but this is unlikely to throw up any interesting structures whilst decreasing the speed of response owing to the need to sort large numbers of similarity coefficients.

The primary use of the searching mechanism described in this section is to allow the chemist a certain degree of browsing ability, allowing him or her to become familiar with the full range of structural types that are present in the file and that are related to the chosen target compound. Perhaps the greatest advantage of this system, albeit an unquantifiable one, is that it enables the retrieval of unexpected, but structurally similar, molecules in response to a query that may act as a powerful spur to the chemist's imagination, suggesting new structural approaches to the problem in hand. Such an exploratory, browsing-like approach to chemical structure searching, together with the substructural ranking procedure described in the previous section, represents a considerable augmentation of the facilities available in current chemical information systems.

REFERENCES

Adamson G.W. and Bawden, D. (1975). A method of structure-activity correlation using Wiswesser Line Notation. J. Chem. Inform. Comput. Sci. 15, 215-220.

Adamson, G.W. and Bawden, D. (1980). Automated additive modelling techniques applied to thermochemical property estimation. J. Chem. Inform. Comput. Sci. 20, 242-246.

Adamson, G.W. and Bush, J.A. (1973). A method for the automatic classification of chemical structures. Inform. Stor. Ret. 9, 561-568.

Adamson, G.W. and Bush, J.A. (1975). A comparison of the performance of some similarity and dissimilarity measures in the automatic classification of chemical structures. J. Chem. Inform. Comput. Sci. 15, 55-58.

Adamson, G.W. and Bush, J.A. (1976). Evaluation of an empirical structure-activity relationship for property prediction in a structurally diverse group of local anaesthetics. J. Chem. Soc. Perkin Trans. I, 168-172.

Bawden, D. (1983). Computerized chemical structure-handling techniques in structure-activity studies and molecular property prediction. J. Chem. Inform. Comput. Sci. 23, 14-22.

Broto, P. Moreau, G. and Vandycke, C. (1984). Molecular structures: perception, autocorrelation descriptor and SAR studies. Eur. J. Med. Chem. 19, 66-70.

Bush, J.A. (1977). Automatic Classification And Structure-Activity Correlation. Ph.D. thesis, University of Sheffield.

Cammarata, A. and Menon, G.K. (1976). Pattern recognition. Classification of therapeutic agents according to pharmacophores. J. Med. Chem. 19, 739-748.

Carhart, R.E., Smith, D.H. and Venkataraghavan, R. (1985). Atom pairs as molecular features in structure-activity studies: definition and application. J. Chem. Inform. Comput. Sci. 25, 64-73.

Chen, B.K., Horvath, C. and Bertino, J.R. (1979). Multivariate analysis and quantitative structure-activity relationships. Inhibition of dihydrofolate reductase and thymidylate synthetase by quinazolines. J. Med. Chem. 22, 483-491.

Cone, M.M., Venkataraghavan, R. and McLafferty, F.W. (1977). Molecular structure comparison program for the identification of maximal common substructures. J. Am. Chem. Soc. 99, 7668-7671.

DiPaolo, T. (1978a). Structure-activity relationships of anaesthetic ethers using molecular connectivity. J. Pharm. Sci. 67, 565-566.

DiPaolo, T. (1978b). Molecular connectivity in

quantitative structure-activity relationship study of anesthetic and toxic activity of aliphatic hydrocarbons, ethers and ketones. J. Pharm. Sci. 67, 566-568.

Gabanyi, Z., Surjan, P.and Naray-Szabo, G. (1982). Application of topological molecular transforms to rational drug design. Eur. J. Med. Chem. 17, 307-311.

Gray, R. (1985). Databases, downloading and proprietary rights. Program 19, 311-321.

Hall, L.J. and Kier, L.B. (1978). Antimicrobial activity of substituted phenyl propyl ethers. J. Pharm. Sci. 67, 1743-1747.

Hansch, C., Gricco, C., Silipo, C. and Vittoria, A. (1977). Quantitative structure-activity relationships of chymotrypsin-ligand interactions. J. Med. Chem. 20, 1420-1435.

Hansch, C. and Yoshimoto, M. (1974) Structure-activity relationships in immunochemistry. 2. Inhibition of complement by benzamidine. J. Med. Chem. 17, 1160-1167.

Harrison, P.J. (1968). A method of cluster analysis and some applications. Appl. Stat. 17, 226-236.

Henry, D.R. and Block, J.H. (1979). Classification of drugs by discriminant analysis using fragment molecular connectivity values. J. Med. Chem. 22, 465-472.

Kier, L.B. and Hall, J.L. (1976). Molecular connectivity. VII. Specific treatment of heteroatoms. J. Pharm. Sci. 65, 1806-1809.

Kier, L.B. and Hall, J.L. (1979). Molecular connectivity analyses of structure influencing chromatographic retention indices. J. Pharm. Sci. 68, 120-122.

Kier. L.B., Murray, W.J. and Hall, L.H. (1975). Molecular connectivity. 4. Relationship to biological activities. J. Med. Chem. 18, 1272-1274.

Lynch, M.F. and Willett, P. (1978). The automatic detection of chemical reaction sites. J. Chem. Inform. Comput. Sci. 18, 154-159.

McGregor, J.J. and Willett, P. (1981). Use of a maximal common subgraph algorithm in the automatic identification of the ostensible bond changes occurring in chemical reactions. J. Chem. Inform. Comput. Sci. 21, 137-140.

Minchin, R.F. and Madsen, B.W. (1980). Inhibition of chlorphentermine binding in rat lung: application of connectivity theory. Life Sci. 27, 1041-1046.

Mortensen, E. (1984). Downloading in online searching: a review of the literature, in Proceedings of the Eighth International Online Information Meeting. Learned Information, Oxford.

Randic, M. and Wilkins, C.L. (1979). Graph theoretical approach to recognition of structural similarity in molecules. J. Chem. Inform. Comput. Sci. 19, 31-16.

Richard, A.J. and Kier, L.B. (1980). Structure-activity analysis of hydrazide monoamine oxidase inhibitors using molecular connectivity. J. Pharm. Sci. 69, 124-126.

Salton, G. and McGill, M.J. (1983). Introduction To Modern Information Retrieval. McGraw-Hill, New York.

Siegal, S. (1956). Nonparametric Statistics. McGraw-Hill, Tokyo.

Sparck Jones, K. (1973). Index term weighting. Inform. Stor. Retr. 9, 619-633.

Van't Riet, B., Kier, L.B. and Elford, H.L. (1980). Structure activity relationships of benzohydroxamic acid inhibitors of ribonucleotide reductase. J. Pharm. Sci. 69, 856-857.

Varkonv, T.H., Shiloach, Y. and Smith, D.H. (1979). Computer-assisted examination of chemical compounds for structural similarities. J. Chem. Inform. Comput. Sci. 19, 104-111.

Wilkins, C.L. and Randic, M. (1980). A graph theoretical approach to structure-property and structure-activity correlations. Theor. Chim. Acta. 58, 45-68.

Wilkins, C.L., Randic, M., Schuster, S.M., Markin, R.S., Steiner, S. and Dorgan, L. (1981). A graph-theoretic approach to quantitative structure-activity/reactivity studies. Anal. Chim. Acta. 133, 637-645.

Willett, P. (1986). Modern Approaches to Chemical Reaction Searching. Gower, Aldershot.

Willett, P. and Winterman, V. (1986). A comparison of some measures for the determination of inter-molecular structural similarity. Quant. Struct. Activ. Relat. 5, 18-25.

Willett, P., Winterman, V. and Bawden, D. (1986). Implementation of nearest neighbour searching in an online chemical structure search system. J. Chem. Inform. Comput. Sci., 26, 36-41.

CHAPTER 4
Clustering in Chemical Information Systems

4.1 A REVIEW OF THE CLUSTERING OF CHEMICAL STRUCTURES

The basic concepts of cluster analysis have been described in Chapter 2. The present chapter starts by reviewing the use of cluster analysis methods in chemical information systems; this is followed by a validation study of the appropriateness of clustering methods for the analysis of sets of chemical structures characterised by fragment descriptors. An extended comparative investigation then follows to determine which types of clustering method are most useful for the classification of chemical structures, and the chapter is completed by examples of the use of cluster analysis techniques in an operational chemical information system.

Cluster analysis as a discipline has become generally feasible only with the widespread availability of computers, and it is thus not surprising that chemical applications of cluster analysis have a fairly recent history. An early contribution is that of Sneath (1966) who described a group average classification of the 20 naturally occurring amino acids based upon physicochemical properties and the presence or absence of a range of structural features. The classification was found to be in fair agreement with intuitive estimates of the structural relationships present in the data set; for example, the long-chain alkylamino acids, leucine, isoleucine and valine, formed a well-marked group, as did the aromatic acids tyrosine, phenylalanine and tryptophan. A principal components analysis identified four major chemical factors in the data set, and correlations of these factors with biological activity data suggested that the methods might be useful for the prediction of biological activity; Sneath noted, however, that such predictions would probably not be of any great accuracy, although better than random estimates.

Shortly after Sneath's paper, Harrison (1968) described an experimental program which had been developed at ICI Pharmaceuticals Division for the identification of clusters in groups of active and inactive compounds. The structures were characterised by binary fragment descriptors, and an analysis of the centres of gravity of the various clusters permitted the identification of those fragments which contributed strongly to activity or inactivity: thus the system might best be considered as an early example of substructural analysis, as described in Chapter 1, in that it provides a means of evaluating the utility of different substructural descriptors. Colman (1968) discussed the use of this program for the classification of 216 streptococci, and found that the clusters formed were similar to those produced by a single linkage clustering, while White and Lewinson (1977) described a modified version of the approach.

The amino acids studied by Sneath formed the initial basis for the important work of Adamson and Bush (1973). This involved simulated property prediction of pI values using a single linkage clustering of the 20 compounds, and their methodology forms the basis for the extended studies reported later in this chapter. Their main contribution was to show that classifications of chemical compounds could be obtained solely upon the basis of fragment substructures similar to those used on a routine basis in chemical structure search systems. In her doctoral thesis, Bush (1977) extended the work to two other data sets, these being 39 structurally diverse molecules with local anaesthetic activity and a collection of 79 penicillin derivatives for which the serum binding had been determined, and studied a variety of fragment descriptor types and similarity measures.

Since these initial studies, several workers have discussed the use of cluster analysis for chemical applications, and this work is now reviewed briefly to provide the context in which our own experimental studies are to be viewed.

A comparison of agglomerative clustering techniques is described by Adamson and Bawden (1981) using a set of 36 benzene derivatives characterised by the frequencies of occurrence of each of 7 substituents, regardless of position. Of the methods tested, the single linkage and complete linkage clusters were the least highly structured, with the single linkage method resulting in an almost featureless classification when unstandardised data was used; Ward's method, on the other hand, resulted in a distinctive pattern of widely separated clusters. Standardisation of fragment occurrence data was found to weight the clustering in favour of the less frequently occurring features, and thus the classifications obtained with standardised data were found to be based primarily upon the type of substituent that was present. The classifications obtained with unstandardised data, conversely, were found to be based in large part upon carbon skeletal information. It may be noted here that the latter form of classification is likely to be of more use when large heterogeneous data sets need to be analysed, rather than the small number of analogues studied by Adamson and Bawden.

Agglomerative methods have also been suggested for the clustering of chemical structures on the basis of activity spectra, i.e., the activities of the compounds in a battery of biological tests. Thus Saggers (1974) discusses the use of single linkage clustering on a small set of compounds with herbicidal activity, although few details of the work are given, while Lewi (1976) describes the use of cluster analysis and principal components analysis for the classification of 40 neuroleptics characterised by 12 assays in rats. A further example is provided by the work of Takahashi et al. (1980) who reported the use of an average linkage method to cluster 29 antibiotics on the basis of antibacterial data. and found that the 6 classes of antibiotic present in the data set could be identified in the resulting dendrogram, given a suitable choice of similarity threshold.

A rather different use of agglomerative clustering has been suggested by Hansch et al. (1973) as a means of selecting substituent groups for Hansch analysis so that the physicochemical properties of each analogue tested in an optimisation program would be as distinct as possible. 90 commonly occurring groups were clustered and a series of thresholds applied to the resulting dendrograms to obtain clusters of substituents. One substituent was then isolated from each such cluster so as to obtain a set for which the corresponding analogues would span the entire physicochemical parameter space. The choice of threshold, and hence the number of clusters identified, could be varied depending upon how many analogues needed to be synthesised. A practical example of the use of this technique is given by Dunn et al. (1976) in the context of an optimisation study of antitumour triazenes.

As has been noted in Chapter 2, the minimum spanning tree (MST) is closely related to the single linkage agglomerative method, and several examples of chemical applications of MST clustering have been reported in the literature. Chu (1974) used a range of pattern recognition and clustering techniques, including MST clustering, to classify a set of 66 structurally diverse molecules as either tranquillisers or sedatives; the structures were characterised by augmented atom fragments, and accurate activity predictions were obtained with all of the procedures that were tested. Ritter and Isenhour (1977) described the use of an MST-based clustering algorithm, related to those described by Zahn (1971), to group 226 chromatographic liquid phases using retention information for 10 reference compounds. Their analysis identified 41 outlier phases with the remaining phases being grouped into 13 clusters, and provided a quantitative basis for the selection of standard stationary phases. Miyashita et al. (1981) studied the MSTs obtained for 62 cephalosporins characterised by antibacterial spectra, and 38 benzodiazepines characterised by 8 physicochemical parameters. They suggested that the longest single path in the MST could be used as a means of obtaining a representative sub-set of a collection of compounds.

A rather different use of hierarchic clustering is evident in the work of Zupan (1980) who describes an heuristic algorithm for the generation of approximate group average hierarchies; a very similar, but more efficient, procedure had been described earlier by Williamson (1974) for the clustering of documents in information retrieval systems. Zupan (1982, 1983) has used the algorithm to increase the efficiency of a search system for infra-red spectra, and it is claimed to result in intuitively sensible groupings of structurally related compounds. However, the algorithm operates by the assignment of new molecules to a partially completed hierarchy, and this means that the classification obtained in any particular computer run will depend crucially upon the exact order in which the file of compounds is processed: Delaney (1981) describes a modified version of the algorithm which is claimed to reduce this problem.

Hierarchic divisive clustering methods seem to have been used very much less than the agglomerative procedures in clustering applications, although Chu (1974) included a monothetic divisive method in his comparison of a range of classification procedures. A further example is provided by the work of Kaufman et al. (1983) and Massart and Kaufman (1983) who used the polythetic procedure described by MacNaughton-Smith et al. (1964) to cluster 53 coals on the basis of their elemental and mineral compositions.

An example of a non-hierarchic clustering method that has been used extensively in analytical chemistry applications is the MASLOC procedure of Kaufman and his co-workers (Massart et al., 1983). This method selects some number, P, of objects from a data set of size N in such a way that the sum of the distances of the other N-P objects to the closest of the selected ones is minimised; each of the N objects is then assigned to one of the clusters, this giving a non-overlapping partition of the data set. An exact solution to the minimisation problem cannot be obtained unless the size of the data set is very small, and thus the procedure adopts 2 heuristics which, either singly or in combination, permit two drastic reductions in computational requirements while still achieving acceptable results in many cases; the first of these heuristics is similar to that suggested by Zhitkov for the classification of images (1970). The procedure has been applied to several problems in analytical chemistry including the characterisation of chromatographic stationary phases (De Clercq et al., 1976), and the classification of archaeological artefacts (Massart et al., 1980). A comparison of the performance of the MASLOC algorithm with several hierarchical agglomerative methods suggested that the MASLOC procedure yielded more useful results in a study of classifications of iron meteorites (Massart et al., 1982).

Rather than grouping previous research on the basis of the clustering procedure that has been adopted, an alternative overview of the work may be obtained by considering the type of attribute that has been used for the characterisation of the compounds under investigation. The first such type, and the main form of representation that is studied in this volume, is the use of substructural descriptors as suggested by Harrison (1968) and by Adamson and Bush (1973). Alternatively, compounds may be clustered on a spectral basis as in the work of Lewi (1976) on spectral mapping, or that of Zupan (1982) on spectroscopic data. Physicochemical parameters may be used as a basis for the clustering of molecules (Miyashita et al., 1982), but also of substituents (Hansch et al., 1973). Finally, most clustering applications in analytical chemistry use types of attribute that are specific to the particular application (Massart and Kaufman, 1983).

Having given a brief outline of previous work in the

field, the main body of this chapter discusses an extended series of experiments which have been carried out to test the utility of several classes of clustering method for the grouping of chemical structures characterised by fragment substructures. The initial studies in Section 4.2 provide a formal validation of the appropriateness of cluster analytic techniques for the grouping of chemical structures, and this is followed in Section 4.3 by systematic comparative experiments using all of the major types of clustering method described in Chapter 2. In both cases, the structures are characterised by a list of the augmented atom fragments contained within them, generally with the frequencies of occurrence of each such substructure. Most of the clustering routines used in these experiments are taken from the CLUSTAN package, a widely used and very flexible set of FORTRAN routines (Wishart, 1978). The small-scale tests provide the basis for the more extended clustering experiments described in Section 4.4 which describe the implementation of clustering procedures in an operational chemical information system.

4.2 THE VALIDITY OF CHEMICAL CLASSIFICATIONS

Dubes and Jain (1979) identify three main types of cluster validity study in their excellent review of the subject. The first of these seeks to determine whether the clustering tendency is such as to suggest that the data is significantly different from random, since it is clearly inappropriate to consider the application of a clustering procedure if this is not the case; these studies of clustering tendency have generally involved the description of randomness in either graph-theoretic (Fillenbaum and Rapaport, 1971; Schultz and Hubert, 1973; Hubert, 1974; Ling, 1975; Ling and Killough, 1976) or geometric terms (Strauss, 1975; Saunders and Funk, 1977). Once it has been shown that a data set does indeed exhibit a non-random clustering tendency, tests may be carried out to determine the extent to which the output from the classification procedure recovers the structure of the data set: this factor is considered in more detail in Section 4.3.2.2 where we evaluate the dendrograms resulting from the use of hierarchical agglomerative clustering methods. Finally, once the overall structure has been confirmed, the validity of individual clusters within a hierarchy or a partition may be of interest (Ling, 1973; Bailey and Dubes, 1982).

The random graph hypothesis is a null hypothesis which is invoked to determine whether any significant predisposition to cluster is present within a data set. Given a data set containing N objects, it is assumed that a symmetric NxN matrix, S, is available, the elements of which, S[I,J], contain the Euclidean distances between each pair of objects I and J. The rank matrix R is another NxN symmetric matrix, the elements of which, R[I,J], contain the rank orders of the corresponding elements in P after they have been sorted into order of increasing distance; the upper portion of R will accordingly contain all integers in the range 1 to N(N-1)/2. The random graph hypothesis is that all [N(N-1)/2]! rank matrices are equally likely, and this null hypothesis may be used to study various characteristics of clusterings. One such characteristic is the minimum number of edges, VMIN, at which a random graph on N nodes becomes connected, i.e. when all of the nodes are contained within a single cluster. Following earlier, and approximate, work by Fillenbaum and Rapoport (1971) and by Schultz and Hubert (1973), Ling (1975) found an accurate method for the enumeration of all connected graphs on N nodes and V edges, and Ling and Killough (1976) used this to calculate tables for the probability of observing specific values of VMIN under the random graph hypothesis given a graph containing N nodes.

The connected subgraphs of a graph may be obtained by setting some threshold distance T and then linking together all pairs of objects for which S[I,J] <= T. The resulting subgraphs correspond to the clusters formed at that threshold by the single linkage clustering method, and Ling and Killough's tables may hence be used by studying single linkage cluster hierarchies. Specifically, a note is made of the rank in R at which all of the objects in a data set become contained within a single cluster; this rank may then be compared with the tables which give values for the probability of observing such a value of VMIN on the assumption that the graph is not a random one. If this probability is greater than some arbitrary threshold value, for which Dubes and Jain (1979) suggest 0.99, then evidence exists for the conclusion that R is significantly different from a random graph, and that some non-random clustering tendency is present in the data set. Thus, the degree of clustering tendency may be tested for simply by determining whether the observed VMIN is large when compared to the distribution of VMIN under the assumption of a random graph.

The reader should note that Ling and Killough's tables do not provide a formal test of statistical significance for the existence of clusters since this would require a meaningful alternative hypothesis embodying some explicit definition of cluster structure. In the absence of such a hypothesis, the primary use of the test is in helping to determine whether clustering techniques are an appropriate means of analysing a particular data set: it is for this purpose that the test is used here.

The experiments involve 14 small sets of chemical compounds, specifically the 16 data sets described in the similarity measure experiments of Chapter 3 with the exception of the two largest groups of compounds, data sets F and P, since the tables provided by Ling and Killough can be used only for data sets containing less than 100 objects. The single linkage classification for each such data set was analysed to identify the level at which all of the members of the data set were contained within a single cluster to determine whether any significant clustering tendency was present. It should be emphasised that although the single linkage is used, the test validates not just this particular clustering method but the clustering tendency that is present in the source data; a random result in the test implies that all clustering methods are inappropriate, and not just the single linkage method.

Random clustering behaviour is observed in only two of the sets of compounds. The first of these is data set I, a set of 27 aliphatic hydrocarbons, ethers and ketones for which the the observed VMIN value is 60, well below the critical value of 88. The second set of compounds, data set K, contains 28 substituted phenyl propyl ethers. The circumstances here are rather different, since the observed VMIN is 355, as against a considerably lower critical value for N=28 of 92, and it might hence be assumed that the data set is indeed clustered. However, a limitation of Ling and Killough's test is that it is based on the point at which the graph becomes fully connected, and thus reflects the clustering tendency at only a single rank in the evolution of the hierarchy; accordingly, the test may be strongly affected by outliers that are far removed from the other members of the data set. The observed value of VMIN=355 for the phenyl propyl ethers arises from just such an outlier, since it joins the hierarchy long after all of the molecules which are totally connected at VMIN=85. This behaviour is shown in the dendrogram on the next page. Accordingly, it is only the presence of a single molecule that causes the compounds to exhibit non-random clustering behaviour, and the data set should thus be regarded as unclustered.

Examples of the validation of clustering tendency in other sets of compounds are given by Willett (1985); this paper also describes the use of the test for analysing sets of substituents for Hansch analysis. Examples of other applications are given by Ling and Killough (1976) and by Dubes and Jain (1979).

4.3 THE EVALUATION OF CLUSTERING METHODS

4.3.1 Introduction

It will be clear from the review of cluster analysis given in Chapter 2 that there are very many different types of clustering method which can be used for the analysis of a set of objects. Several of these methods have been used at one time or another for the clustering of chemical compounds, as described in Section 4.1, and an investigator who wishes to use cluster analysis methods is thus faced with the problem of deciding which method to use. This is a difficult problem since comparative tests of clustering methods with a given data set may well result in the production of widely disparate classifications, although this may be helpful in some cases since it provides alternative views of the groupings present within the data. It would, therefore, be generally

Single linkage classification of 28 phenyl propyl ethers

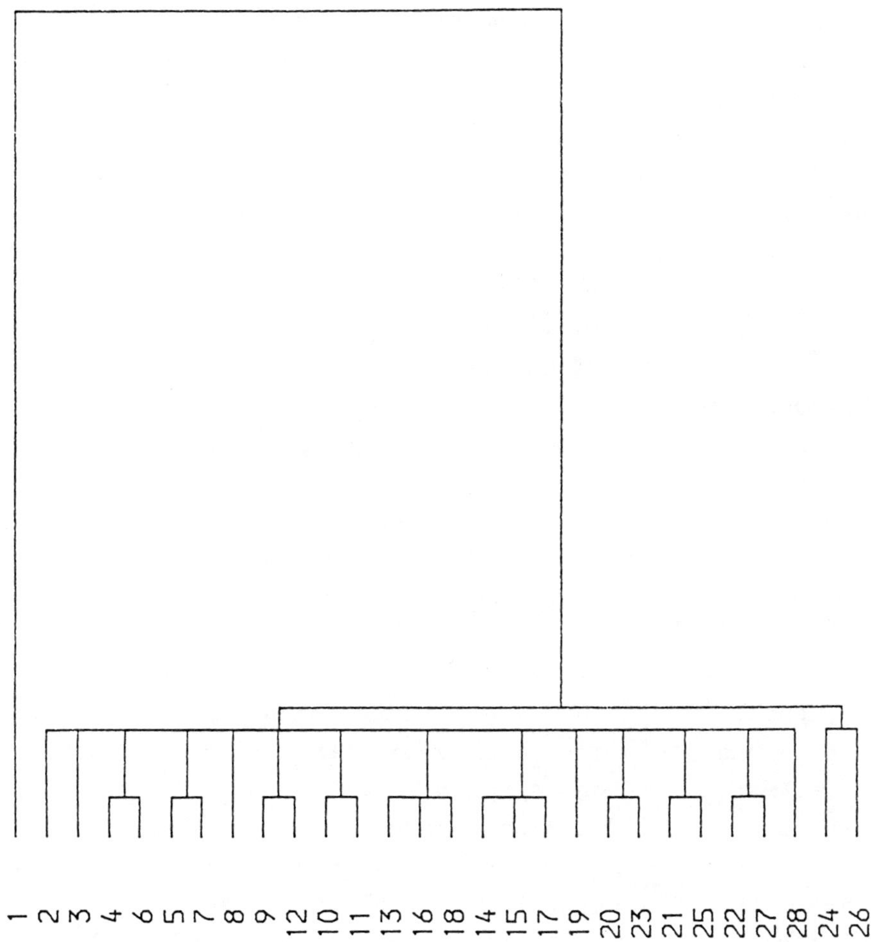

helpful if some broad guidelines were available to assist in the selection of a method, or methods, for use in this particular application area. To this end, an extended comparative survey has been undertaken of the use of a range of clustering methods for the classification of chemical compounds characterised by fragment descriptors.

There are many ways in which a classification may be evaluated. In the context of operational chemical information systems, an important feature of a clustering method is the extent to which it can group together molecules with comparable properties, i.e., the extent to which a classification based on substructural characteristics mirrors similarities in physical, chemical and biological properties. Thus a 'good' classification will be one in which molecules with similar properties are grouped together in a cluster, while being well separated from other clusters containing molecules with different characteristics. Such an approach is, of course, comparable to the evaluation criterion used in the similarity measure tests of Sections 3.2 and 4.2, and it forms the basis for all of the experiments in the present section.

The evaluation involves the data sets used in Section 3.2, specifically the 14 which have been shown to exhibit a non-random clustering tendency. Each of these data sets is clustered using some clustering method, and the resulting classification then used for simulated property prediction by means of the methodology of Adamson and Bush (1973). The property value for each molecule, I, within a data set is assumed to be unknown, and the classification resulting from the use of some clustering method is scanned to identify the cluster containing I; the <u>predicted</u> property value for I, P[I], is then set equal to the mean of the actual property values of the other compounds in that cluster. Such a 'leave-one-out' procedure results in the calculation of a P[I] value for each of the N structures in the data set, and an overall figure of merit for the classification is then obtained by calculating the product moment correlation coefficient between the sets of N actual and predicted values. The most generally useful clustering methods will thus be those which give high correlation coefficients across as wide a range of data sets as possible.

While intuitively reasonable, such a methodology presents problems when hierarchical clustering methods are used, since an individual compound occurs in a whole series of clusters as the agglomeration, or division, proceeds, including the single cluster that represents the entire data set. A partition can be obtained by applying a similarity threshold to the hierarchy, and using the clusters present at that threshold as the basis for the prediction of property values. However, this would involve the specification of an appropriate threshold, and might result in a partition in which some members of the data set are still unclustered. Accordingly, the hierarchic clustering experiments considered in Sections 4.3.2.1, 4.3.3 and 4.3.4 are evaluated using the concept of a bottom level cluster, where the bottom level cluster for the I'th compound is that cluster which the I'th compound joins when it first becomes linked into the cluster hierarchy. This technique was originally suggested by Adamson and Bush (1973) in their early studies of chemical classifications.

The evaluation of clustering methods by property prediction tests emphasises the effectiveness of the various clustering methods, without regard to the efficiency with which they may be implemented, i.e., to the computational resources required for the generation of a classification. While a computationally demanding clustering method may be appropriate for the analysis of small numbers of molecules, such as the data sets used in these property prediction experiments, it will not be applicable to large-scale clustering. Accordingly, one of the main aims of the present comparative study is to determine the extent of the trade-off between efficiency and effectiveness, i.e. the extent to which there is a deterioration in predictive performance as computational requirements lessen, so as to enable the identification of a clustering method which most successfully reconciles these two conflicting characteristics.

4.3.2 Hierarchic agglomerative clustering methods

4.3.2.1 Property prediction

The predictive tests with the hierarchic agglomerative methods used ten of the data sets of Chapter 3, these being sets A, B, D, F, H, J, L, N, O and P. Each set of compounds was clustered using the single linkage, complete linkage, group average and Ward methods, with the Euclidean distance being employed for the calculation of the inter-molecular similarities. The results obtained from the use of a weighted representation, i.e. one in which each molecule is represented by a list of the frequencies of occurrence of each fragment present, are listed in Table 4.1. A starred value represents a correlation coefficient that is not statistically significant at the 0.05 level of statistical significance.

An inspection of the results reveals two main types of behaviour. These are when all four methods give comparable results, e.g., the benzamidine and N-acyl ester data sets, and when the single linkage method gives results that are much worse than those for the complete linkage, group average and Ward methods; examples of such behaviour include the amino acid, local anaesthetic and tadpole narcosis data sets. The conclusion that the single linkage method is noticeably inferior to the other three methods is supported by several recent studies of the use of hierarchic agglomerative clustering methods for automatic document retrieval (Griffiths et al., 1984; Vorhees, 1985; Griffiths et al., 1986). This work has involved the searching of queries against the bottom level clusters of

Data set	Single linkage	Complete linkage	Group average	Ward's method
A	0.54	0.94	0.93	0.94
B	0.64	0.87	0.86	0.86
D	*	0.54	0.34	0.35
F	0.85	0.86	0.83	0.87
H	0.85	0.88	0.87	0.87
J	0.67	0.73	0.71	0.77
L	0.69	0.78	0.78	0.79
N	0.55	0.63	0.64	0.65
O	0.62	0.77	0.82	0.78
P	0.83	0.93	0.92	0.86

Table 4.1

Hierarchic agglomerative clustering using weighted fragment descriptions

document hierarchies. and has emphasised the generally poor level of retrieval effectiveness that can be obtained from searching the single linkage hierarchies. Griffiths et al. (1986) provide evidence to suggest that there is very little difference in the membership of the bottom level clusters arising from the complete linkage, group average and Ward methods, and that these clusters are substantially different from the single linkage bottom level clusters. It is thus hardly surprising that the three methods give comparable levels of performance in document retrieval and property prediction tests.

The generally poor performance of the single linkage classifications is explained in large part by an inspection of the corresponding dendrograms. The single linkage method gives much less highly structured classifications than the other three methods, with very few distinct clusters being identified, a phenomenon which is referred to as chaining. Accordingly, because the predicted property value for a structure is the average of the values for all of the other structures in a cluster, the predictive performance is likely to be poor if only a very few clusters are identified by the classification procedure. Dendrogram structures are discussed in detail in Section 4.3.2.2. It should be noted that the single linkage method does have the advantage that it can be implemented very efficiently, and on a much larger scale than can the other three methods tested here (Sibson, 1973; Willett, 1984a).

Many applications of clustering involve the standardisation of attributes prior to clustering, so as to ensure that all of the attributes contribute equally to the measurement of inter-object similarity. While such a procedure is entirely appropriate when the individual attributes are measured on different types of scale, e.g., lipophilicity and molar refractivity in Hansch analysis, its use is more debatable when all of the attributes are of the same sort and are measured on the same scale. This is, of course, just the case with the data sets studied here, where each molecule is represented in the data matrix by the frequencies of occurrence of individual fragment substructures. To test the effect of standardisation on predictive performance, the fragment occurrences were standardised to zero mean and unit variance, as described in Section 2.2.1, the molecules clustered using the four methods, and the classifications used for property prediction. The resulting correlation coefficients are shown below, and it will be seen that they differ substantially from the results obtained with the unstandardised data.

Data set	Single linkage	Complete linkage	Group average	Ward's method
A	0.50	0.39	0.40	0.38
B	0.36	*	0.49	0.39
D	*	0.52	*	0.55
F	0.65	0.66	0.63	0.69
H	0.70	0.82	0.80	0.85
J	0.69	0.65	0.68	0.71
L	0.75	0.79	0.76	0.77
N	0.52	0.65	0.65	0.65
O	0.56	0.66	0.60	0.72
P	0.89	0.93	0.94	0.94

Table 4.2
Hierarchic agglomerative clustering using standardised fragment descriptions

Predictive performance is sometimes lower when structurally divergent data sets are tested, e.g., the amino acid, tadpole narcosis and local anaesthetic data. With structurally homogeneous data sets, conversely, the correlations are generally comparable with those obtained with unstandardised data, e.g., the N-acyl esters, penicillins and quinazolines. The reason for this behaviour is that the standardisation of fragment occurrences corresponds to reducing the influence of the frequently occurring carbon-based fragments, and to emphasising the less frequently occurring features such as heteroatoms and multiple bonds. It is thus possible that predictive ability will be lessened when the carbon skeleton plays an important part in discriminating between structures, since important structural information will have been eliminated in the course of the standardisation. With sets of analogues, however, the reduction in the importance of the carbon skeletal information resulting from standardisation will have little or no effect on predictive performance since all of the compounds in the data set share a common skeleton. Similar comments apply to properties that are strongly dependent upon the functionality that is present, e.g., molar refractivity and heats of vapourisation.

A more drastic form of standardisation is the use of binary, i.e., presence or absence, data. The correlation coefficients obtained from using binary fragment occurrence information are shown in Table 4.3. The main differences from the previous standardised results are a noticeable increase in the number of non-significant correlations obtained, and a further lessening in overall predictive performance.

Data set	Single linkage	Complete linkage	Group average	Ward's method
A	*	*	*	*
B	0.35	0.55	0.64	0.56
D	*	*	*	*
F	0.79	0.84	0.85	0.83
H	0.84	0.90	0.88	0.88
J	0.62	0.59	0.56	0.59
L	0.69	0.80	0.79	0.80
N	0.38	0.35	0.34	0.34
O	0.34	0.65	0.54	0.66
P	0.86	0.86	0.86	0.85

Table 4.3
Hierarchic agglomerative clustering using binary fragment descriptions

From these three sets of experiments, it seems reasonable to conclude that the weighted molecular characterisations give the most generally useful classifications for property prediction, a result that is in agreement with the findings of the similarity measure tests in Section 3.2. Accordingly, most of the subsequent clustering experiments reported below use unstandardised fragment information. Binary data is used in some cases, either because the clustering method requires it, as with the monothetic divisive procedures of Section 4.3.3, or because frequency data is not available, as with the large-scale experiments reported in Section 4.4.

4.3.2.2 Evaluation of dendrograms

In view of the widespread use of hierarchic agglomerative

clustering methods, further experiments have been undertaken to compare the classifications arising from the use of such methods. This section considers three features of the dendrograms produced by the use of the four methods considered above; as described in Chapter 2, a dendrogram is a graphical representation of a hierarchic classification, showing which molecules and clusters of molecules are fused together and in what order. The three features are the distortions imposed upon an input similarity matrix when a clustering method is applied to it, the degree of similarity between the dendrograms produced by different clustering methods, and the degree of hierarchic structure in the dendrograms. Although none of these features is related directly to the ability of the clustering methods to cluster together structurally related compounds, the experiments provide a useful means of characterising other features of the methods that should be taken into consideration when a clustering method is to be selected for some particular application.

The inter-object similarity data, as reflected in an input similarity matrix, becomes altered, or distorted, by a clustering method during the generation of the cluster hierarchy. This fact suggests an obvious means for the comparison of different clusterings of the same data sets since a clustering method that imposes only a small degree of distortion on the matrix may be thought of as identifying more natural clusters than one which results in a cluster structure that is not actually present in the similarity matrix. It should be noted that such a viewpoint is somewhat controversial. Thus Clifford and Stephenson (1975) and Williams and Clifford (1971) have suggested that distortion of the similarity matrix is not necessarily to be avoided, and that a clustering method should attempt to identify groupings that are more intense than those in the similarity matrix.

Quantitative measures of distortion may be obtained from comparing the set of inter-object similarities comprising an input similarity matrix with the corresponding set of similarities as defined by the output dendrogram. Such measures, which are reviewed by Cormack (1971) and by Rohlf (1974), may also be used to estimate the degree of distortion involved in moving from one dendrogram to another, i.e., the overall degree of similarity between the dendrograms corresponding to different classifications.

The best established measure is probably the cophenetic correlation coefficient, or CPCC (Sokal and Rohlf, 1962). Writing S[I,J] for the inter-molecular similarity for a pair of compounds I and J, and T[I,J] for the similarity level at which they are first clustered together in a single cluster in the hierarchy, the CPCC is simply the product moment correlation coefficient between the sets of N(N-1)/2 S[I,J] and T[I,J] values; thus, the CPCC is scale invariant and measures the extent to which there is a linear relationship between the two sets of similarities. Although it is widely used, Faris (1969) has made several criticisms of the CPCC. He found that the group average method would always tend to give a high value for the measure, because of the clustering criterion used in this method, and questioned the validity of the basic concept by showing that clustering methods that had been designed to optimise the CPCC resulted in dendrogram inversions.

Jardine and Sibson (1971) considered a wide range of measures that may be used for the comparison of dendrograms with similarity matrices and with each other. The dendrogram-matrix distortion measure used here, D, is defined by MAX(¦S[I,J]-T[I,J]¦)/MAX(S[I,J]). When pairs of different dendrograms have to be compared, the two sets of coefficients S[I,J] and T[I,J] need to be normalised. In this case, Jardine and Sibson define D to be MAX¦(S[I,J]/MAXSIJ)-(T[I,J]/MAXTIJ)¦, where MAXSIJ is the largest S[I,J] value and similarly for MAXTIJ. In this case, of course, S[I,J] corresponds to the similarity level at which two compounds first cluster together in one of the dendrograms, rather than the similarity within the similarity matrix. For both types of measure, a value of zero implies complete agreement between the two sets of values S[I,J] and T[I,J].

The CPCC and D measures have been applied to the single linkage, complete linkage, group average and Ward hierarchies for the fourteen valid data sets. The results of these experiments are presented in Table 4.4, which gives the mean and standard deviation for each of the measures when averaged over the 14 data sets.

	CPCC		D	
Single linkage	0.69	0.06	0.72	0.05
Complete linkage	0.69	0.04	0.83	0.05
Group average	0.80	0.04	0.48	0.04
Ward's method	0.58	0.04	3.43	0.83

Table 4.4

Similarity matrix-dendrogram comparisons

	CPCC		D	
Single linkage-complete linkage	0.51	0.09	0.66	0.08
Single linkage-group average	0.79	0.07	0.43	0.07
Single linkage-Ward's method	0.39	0.08	0.83	0.05
Complete linkage-group average	0.77	0.06	0.50	0.07
Complete linkage-Ward's method	0.79	0.05	0.61	0.08
Group average-Ward's method	0.62	0.05	0.71	0.04

Table 4.5
Dendrogram-dendrogram comparisons

Taking the CPCC results first, Faris (1969) suggested that group average would always be expected to give high values for the CPCC, and this is certainly the case with the data sets studied here since the observed CPCC value for this method is considerably greater, i.e., a lower degree of distortion, than for the other three methods. The single linkage and complete linkage methods give results which are almost identical to each other. Ward's method results in a low mean correlation, implying a greater distortion of the input similarity matrix during the operation of the clustering method. With the D figures, the smaller the degree of distortion, the smaller the tabulated value. Here, the group average method again performs well, but the single linkage method now gives a lower degree of distortion than does complete linkage; Ward's method results in by far the largest distortion values.

Rather less consistent behaviour is observed when the inter-dendrogram comparisons are considered. The single linkage classifications are noticeably different from the complete linkage and Ward classifications, but similar to those resulting from the use of the group average method. Ward's method is most similar to complete linkage which is, in its turn, comparable with group average.

The coefficient of Jardine and Sibson used in these experiments is just one of several distortion measures that they proposed. All of these measures are similar in character, and our tests have shown that they all give broadly comparable results: accordingly we have listed the results for only one of them(Table 4.5).

An obvious qualitative means of evaluating different clustering methods is to look at the dendrograms produced to see the amount of structure that has been identified within a given data set. More quantitative approaches to the characterisation of dendrograms have been advocated by several workers (Phipps, 1971; Faris, 1973; Waterman and Smith, 1978; Fowlkes and Mallow, 1983), in particular by Murtagh (1983, 1984) who has defined several coefficients of hierarchic structure and applied them to a wide range of hierarchic classifications. Two of his coefficients are studied here as follows.

Let NBAL be the number of non-terminal clusters in a binary hierarchy, the filial clusters of which differ in size by at most one compound; then the coefficient of hierarchic structure S_1 is defined as (N-1-NBAL)/(N-3), where N is the number of compounds in the collection.

The second coefficient, S_2, considers the numbers of non-singleton clusters at each of the N-1 levels in a hierarchy of N compounds. Let M[J] be the number of such clusters after J fusions, so that some hierarchy H may be characterised by the set of numbers

M[1],M[2]..M[J]..M[N-1].

Two special types of hierarchy may be specified. The first of these, HS, is defined to be as balanced and symmetric as possible, i.e., one in which the N compounds first fuse to give N/2 doubleton clusters, which in turn fuse to give N/4 four-compound clusters, etc. etc. The second special hierarchy, HA, is defined as that which is asymmetric and unbalanced as possible, i.e., one in which the hierarchy is built up by the addition of individual compounds to a single cluster. These two types of extreme behaviour are illustrated on the next page in Figure 4.1. The coefficient of hierarchic structure S_2 is defined by

$$\frac{\sum(MS[J]-M[J])}{\sum(MS[J]-MA[J])}$$

where the summation is over all of the N-1 levels in the hierarchy and where MS and MA are the sequences of numbers of non-singletons for the limiting case hierarchies HS and HA. Both of the coefficients S_1 and S_2 give values in the range from zero to unity, with low values corresponding to balanced or symmetric hierarchies; a more detailed discussion is given by Murtagh in his original papers.

Balanced hierarchies allow the production of useful partitions if a similarity threshold is applied to the hierarchy, and will also facilitate the efficiency of strategies that involve searching the hierarchy from the root of the tree. Such strategies are of considerable theoretical importance in computer science, and have been considered for the organisation of spectral data files by Delaney (1981) and by Zupan (1980). In addition, a clustering method that produces a symmetric hierarchy containing many small clusters is more likely to give useful predictive results than one which produces a few, highly chained clusters containing large numbers of molecules, a fact which has been noted in the discussion

HS

HA

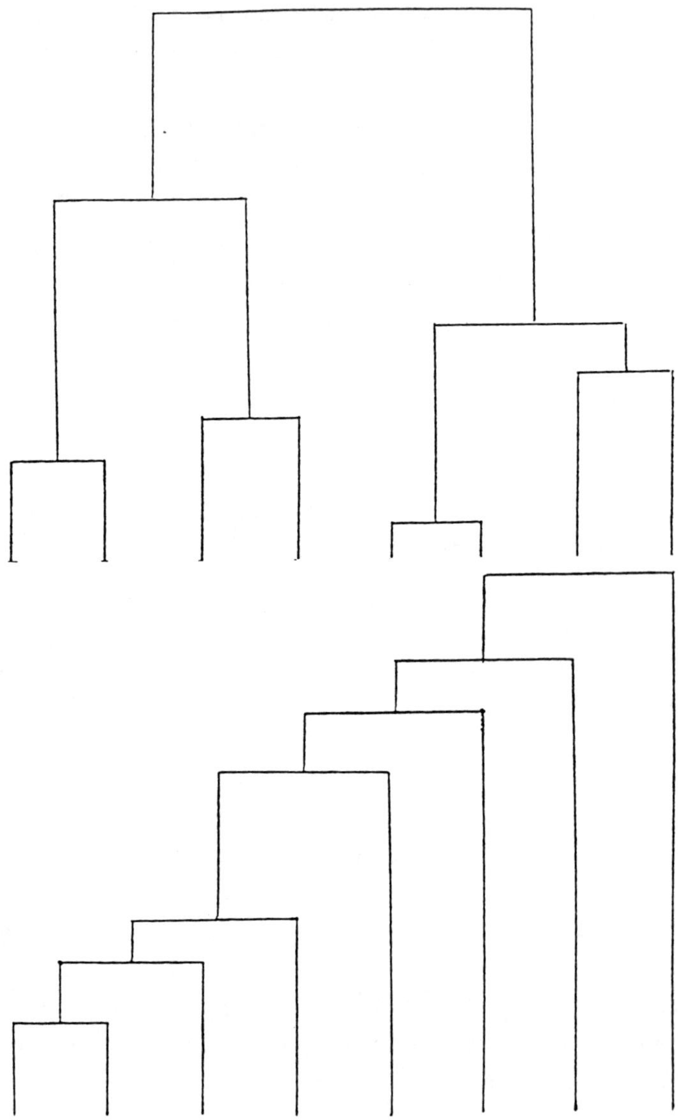

Figure 4.1

of the prediction experiments above.

Examples of dendrograms for the 20 naturally occurring aminoacids are shown on the following pages in Figures 4.2 (a) to (d). Taking the four methods in turn, it will be seen that single linkage, Figure 4.2 (a), produces a relatively small number of large straggly clusters, the chaining phenomenon mentioned earlier; indeed, this dendrogram suggests the existence of a single basic cluster with individual compounds, or small groups of compounds, becoming detached as one moves down the hierarchy. In contrast, the very strict similarity requirements for cluster formation in the complete linkage method, Figure 4.2 (b), results in very many small clusters; group average, Figure 4.2 (c), comes midway between these two extremes, and there is thus not only less evidence of chaining but also a wide range of cluster sizes. Ward's method, Figure 4.2 (d), results in a highly structured dendrogram which is very similar to that resulting from the use of the complete linkage method. Comparable behaviour is seen for the other data sets that were tested, as with the 34 tadpole narcotics illustrated in Figures 4.3 (a) to (d).

The results of the dendrogram structure experiments are presented in Table 4.6 below, which gives means and standard deviations for the two types of measure averaged over all of the data sets.

Single linkage classification of
20 naturally occurring aminoacids

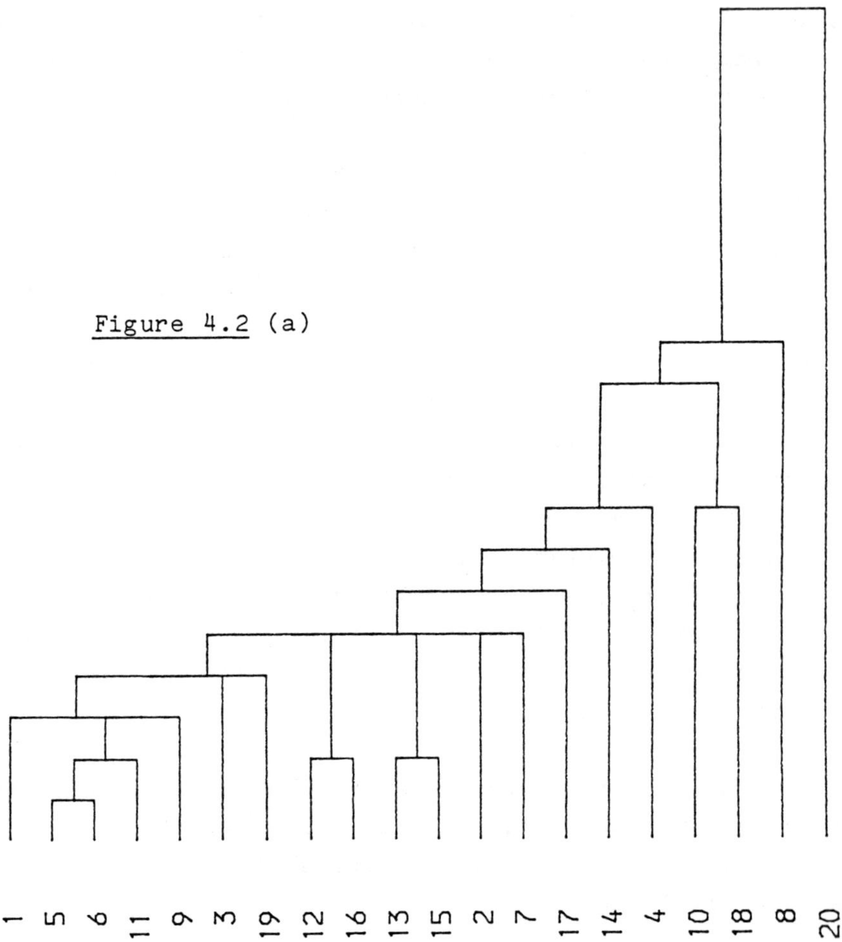

Figure 4.2 (a)

Complete linkage classification of
20 naturally occurring aminoacids

Figure 4.2 (b)

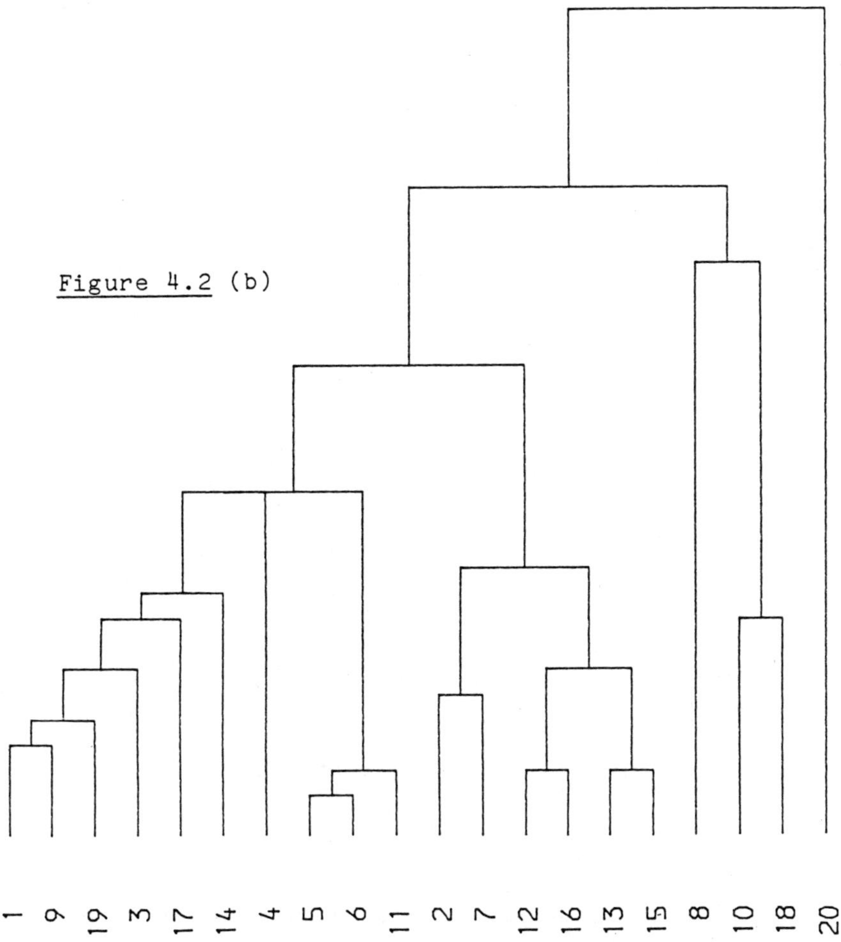

Group average classification of
20 naturally occurring aminoacids

Figure 4.2 (c)

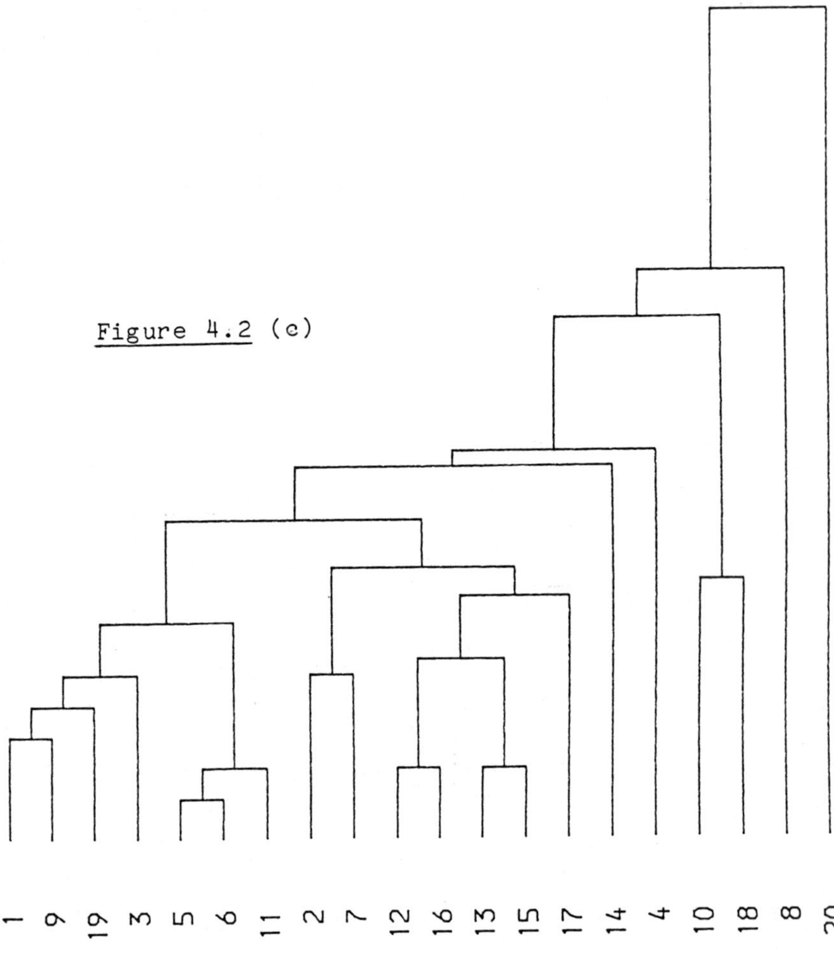

Ward classification of 20
naturally occurring aminoacids

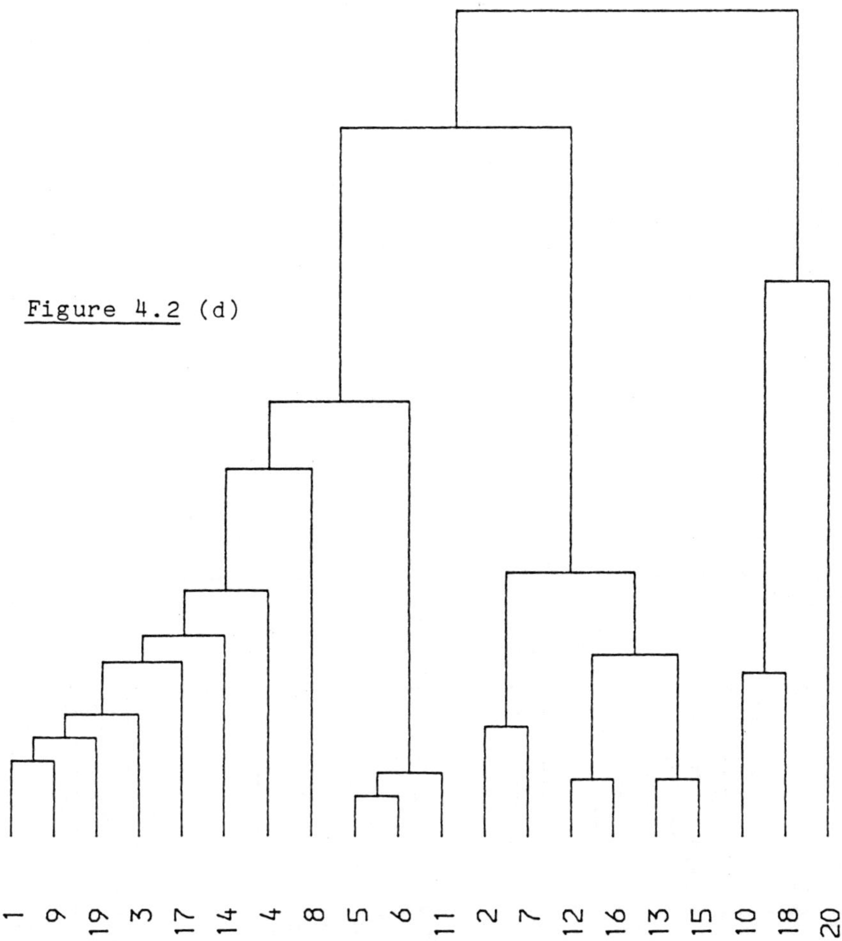

Figure 4.2 (d)

Single linkage classification of 34 tadpole narcotics

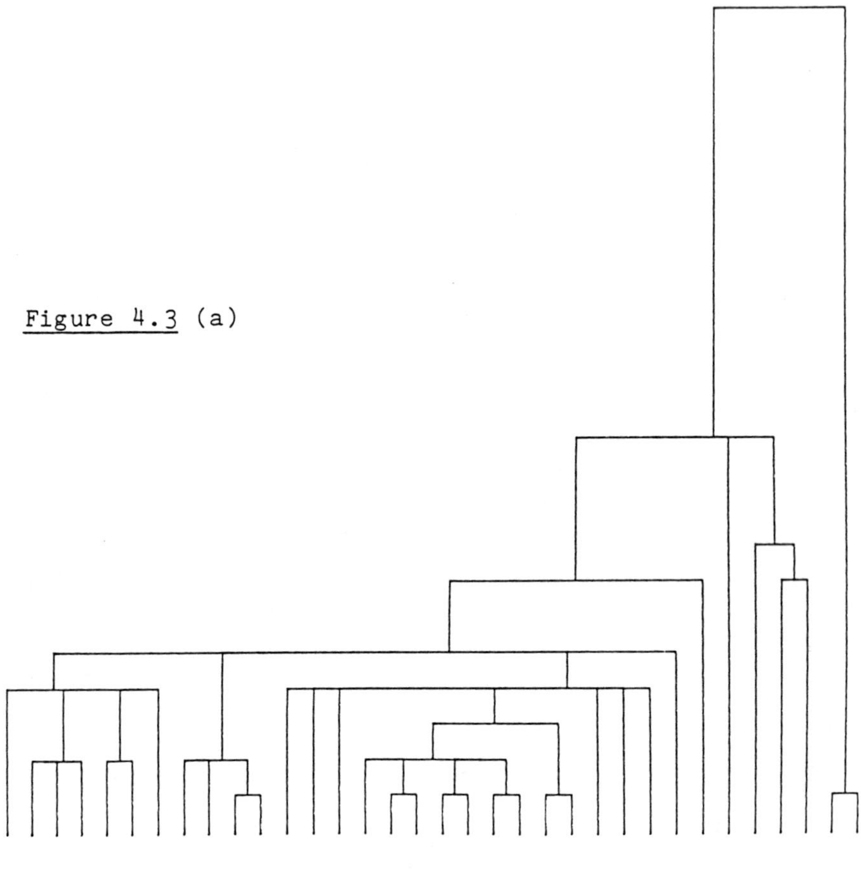

Figure 4.3 (a)

Complete linkage classification of 34 tadpole narcotics

Figure 4.3 (b)

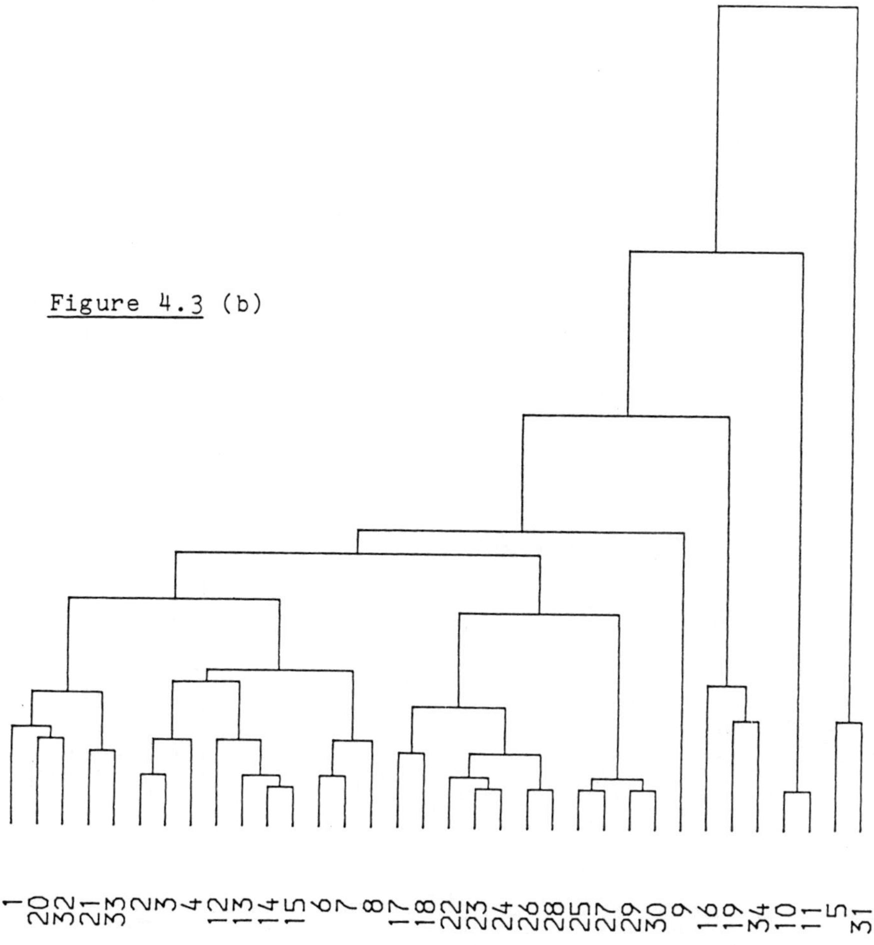

Group average classification of 34 tadpole narcotics

Figure 4.3 (c)

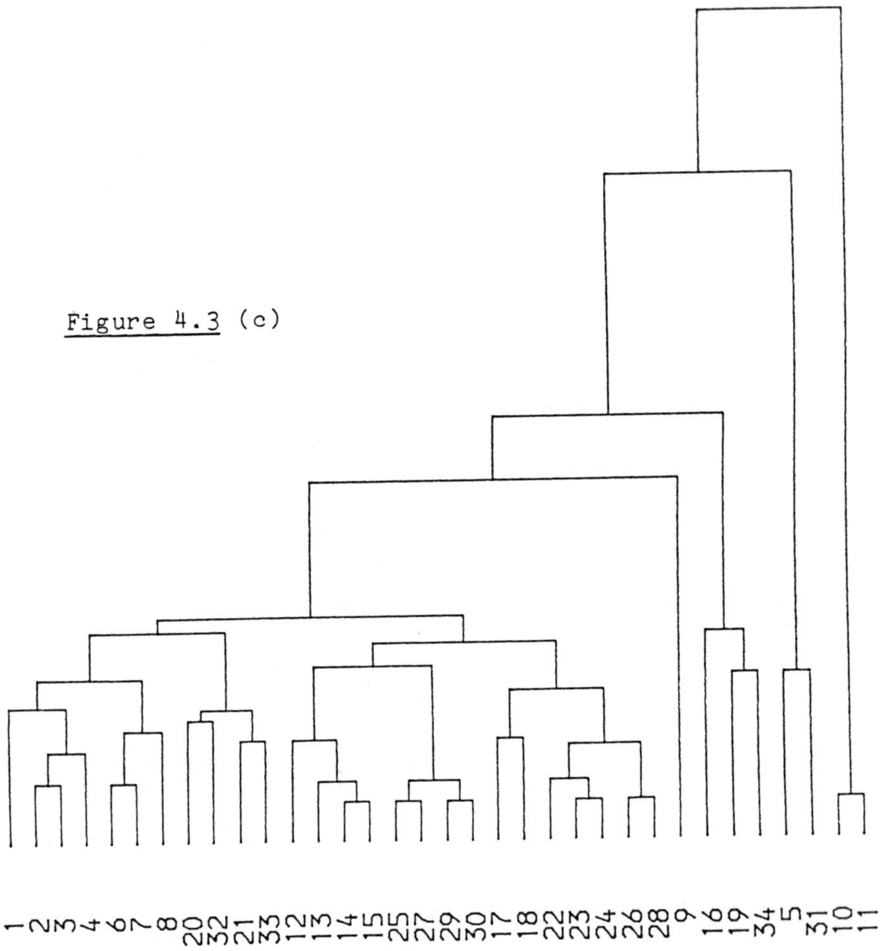

Ward classification of 34 tadpole narcotics

Figure 4.3 (d)

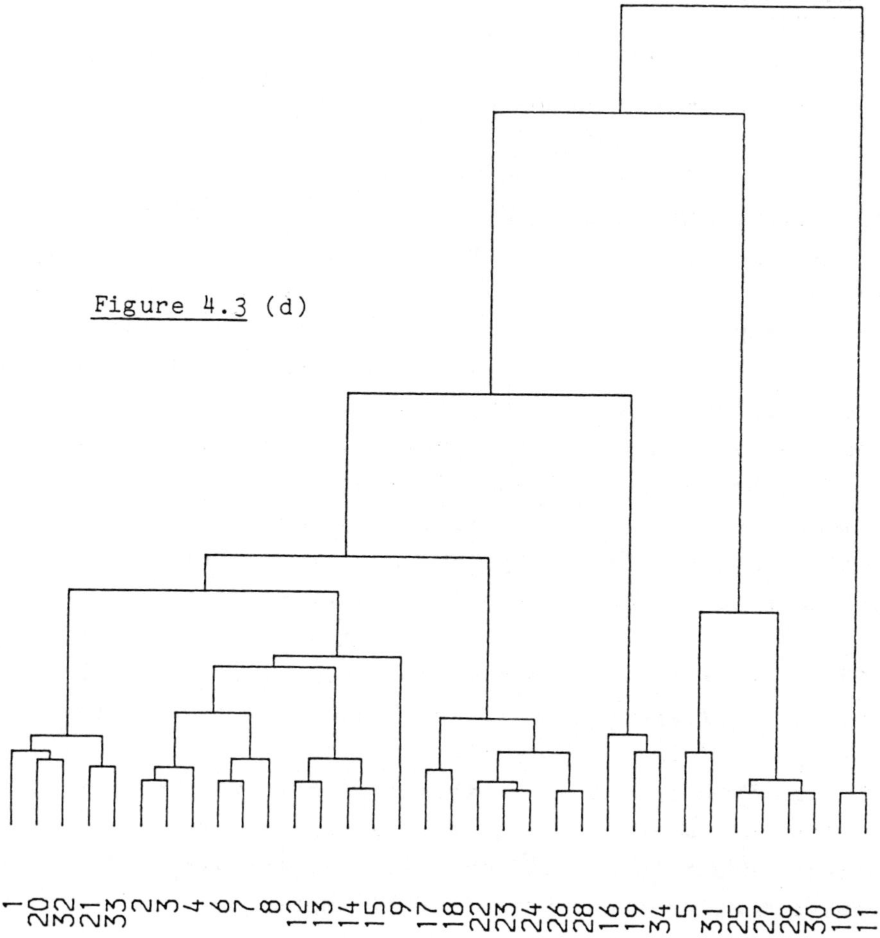

	S_1		S_2	
Single linkage	0.58	0.04	0.65	0.05
Complete linkage	0.39	0.03	0.35	0.04
Group average	0.41	0.03	0.45	0.04
Ward's method	0.37	0.02	0.28	0.03

Table 4.6

Dendrogram structure

The figures emphasise the unbalanced nature of the single linkage classifications, this being in agreement with Murtagh's earlier work (1983, 1984). The coefficient values from the other three procedures are much more similar to each other, with the Ward method giving the lowest values, and with complete linkage the next lowest; this behaviour, which is observed consistently across all of the data sets, is discussed by Murtagh (1983). Of the methods considered here, Ward's method is that most akin to the symmetric hierarchy, HS, and single linkage is that most akin to the asymmetric hierarchy, HA.

The three types of experiments reported in this section consider different characteristics of the dendrograms resulting from hierarchic agglomerative clustering methods. Even so, it is relatively simple to summarise the results by stating that the single linkage and Ward methods result in classifications that are very different from each other, with the other two methods giving results that are intermediate between these two extremes.

Taking the experiments of Section 4.3.2.1 into account, it is suggested that an investigator who wishes to use hierarchic agglomerative clustering methods with sets of chemical structures would be well advised to use at least the single linkage and Ward methods for the analysis.

4.3.3 Monothetic divisive clustering methods

A hierarchic classification of a data set may also be obtained using a divisive strategy in which the classification is built up by the progressive sub-division of an initial single cluster into smaller and smaller sub-clusters. The effectiveness of such methods, specifically the four monothetic methods described in Chapter 2 (minimum error sum of squares (MSS), divisive information analysis (DIS), association analysis (AA) and group analysis (GA)), was tested in a series of experiments comparable to those described in Section 4.3.2.1. The only difference from the agglomerative experiments is that the fragment vectors characterising each structure within a data set are binary in nature, and do not include the number of fragment occurrences, this difference arising from the monothetic nature of the methods. The correlation coefficients between the observed and predicted property values are shown below in Table 4.7.

The first comment that should be made about these results is the fact that no single method performs consistently better or consistently worse than the others, with all yielding acceptable results in some cases. In particular, the computationally efficient group analysis method is not noticeably inferior to the other more time-consuming procedures; however, a limitation of this method, which causes problems of interpretation, is that the distances at which the divisions take place are not monotonically decreasing, causing problems of interpretation when the dendrograms describing the classifications need to be inspected.

Data	MSS	DIA	AA	GA
A	0.54	0.54	*	*
B	0.59	0.67	0.59	0.79
D	*	*	*	*
F	0.86	0.85	0.87	0.85
H	0.86	0.84	0.87	0.88
J	0.67	0.66	0.69	0.67
L	0.78	0.78	0.46	0.40
N	0.36	0.36	0.29	0.33
O	0.51	0.51	0.64	0.73
P	0.84	0.86	0.86	0.81

Table 4.7

Hierarchic monothetic divisive clustering

The absolute magnitudes of the coefficients in the table vary considerably from one data set to another. Some sets of compounds, such as the benzamidines or the N-acyl esters, exhibit uniformly high correlations, while others, such as the amino acids and toxic ethers, give correlations that are not statistically significant. Many of the better correlations are obtained with the sets of analogues in which structural variation is restricted to a limited part of the basic parent compound. The compounds in such a data set differ only in the substituents present, and the monothetic classifications might be expected to be able to reproduce this structure quite accurately; indeed, for the benzamidines, N-acyl esters and quinazolines, the correlations obtained here are comparable with those obtained from the use of the hierarchic agglomerative methods, even using weighted occurrence data. However, a monothetic procedure might be expected to give a good correlation with a structurally disparate set of molecules only if it is possible to identify as the divisive criteria those fragment types which influence the chosen property most strongly. It is thus hardly surprising that the correlations obtained with the heterogeneous data sets are often noticeably less than those obtained with the agglomerative methods, which are polythetic in nature and which would thus seem to be much better suited to such data. Additional results are presented by Rubin and Willett (1983).

4.3.4 Hierarchic relocation clustering methods

The same ten data sets have also been studied using the relocation clustering routine in the CLUSTAN package. This operates by means of a hill-climbing assignment procedure, as described in Section 2.4.2, and also involves a hierarchic routine which is called at the end of each iteration when the relocations have been completed. The routine identifies the most similar pair of clusters and then merges them so that the number of clusters in the partition is reduced by one. It is thus possible to produce a hierarchic classification, despite the fact that relocation methods are normally used for the generation of non-hierarchic classifications. The CLUSTAN package provides seven different definitions of compound-to-cluster and cluster-to-cluster similarity measure. The results reported here are for just two of these measures, these being the Euclidean distance, and what the CLUSTAN manual (Wishart, 1978) describes as the size difference. The Euclidean distance is chosen as it is probably the most widely used measure in the literature of relocation clustering, and to allow comparisons with the agglomerative results reported earlier which all use the Euclidean distance for the calculation of the inter-molecular similarity matrix. Moreover, Willett (1984b) shows that the Euclidean distance gives property prediction results that are comparable with, or superior to, those obtained from five other similarity measures that he tested. All of these measures, which are of varying degrees of complexity, take account of the chemical nature of the fragments characterising each molecule during the calculation of the similarities or distances. The size difference, conversely, takes no account of such information, and instead focusses upon the numbers of fragments present. Specifically, let D[I,J] be an element of the data matrix D containing the frequency of occurrence of the J'th fragment in the I'th structure, and let MD[P,J] be the mean frequency for the J'th fragment in the compounds comprising the P'th cluster. Let MMD[P] be the mean of these scores for the P'th cluster, when averaged over all of the fragment types present. Then the squared size difference between the P'th and Q'th clusters is defined as (MMD[P]-MMD[Q]). The size difference can be regarded as a measure that compares the overall bulk of two objects, i.e., clusters or individual compounds, and is thus analogous to some of the steric and bulk physicochemical parameters that are used in Hansch analysis.

As was noted in Chapter 2, two problems associated with the use of relocation need to be taken into account in any practical use of this approach to clustering. Firstly, the final classification will depend upon the initial assignment of compounds to clusters, with the possibility that different initial assignments may result in substantially different final classifications. Secondly, the operation of the procedure is dependent upon the order in which the set of molecules is processed, owing to the serial nature of the relocation procedure. These two factors thus imply that a better classification, in some sense, might be obtained from an alternative initial partition, or from a re-ordering of the data file. The former problem may be eliminated by adopting some non-random selection procedure to specify the initial partition. The procedure used here derives from the work of Dattola (1971) and involves summing the frequencies of occurrence for each of the fragments in a data set, and then selecting as the initial cluster centres those molecules with the highest sums of fragment frequencies; the initial partition is then obtained by assigning the remaining molecules to the most similar cluster centre. The extent to which the classifications are dependent upon the order of processing is studied by taking five different orderings of each set of compounds and calculating the mean and standard deviation of the correlations. The typical bottom-level cluster in a hierarchic classification contains just a pair of compounds, and thus the initial number of cluster centres in the relocation experiments was set equal to N/2 so as to allow a comparison with the results listed in Sections 4.3.2.1 and 4.3.3.

The results of the property prediction tests are shown in Table 4.8 and it will be seen that the standard deviations are generally quite small, except when Euclidean distance is used with the quinazolines, and thus the order in which the compounds are processed need not seriously affect predictive ability. It may be noted that comparable mean coefficients, but noticeably larger standard deviations, are observed if random initial partitions are used (Willett, 1984b); accordingly, the use of a fixed initial partition provides a simple and effective means of eliminating this undesirable characteristic of many relocation procedures.

Data set	Euclidean distance		Size difference	
A	0.93	0.01	*	
B	0.76	0.03	0.94	0.00
D	0.58	0.05	0.87	0.08
F	0.84	0.01	0.89	0.00
H	0.93	0.00	0.89	0.01
J	0.75	0.02	0.48	0.06
L	0.75	0.09	0.39	0.02
N	0.59	0.02	0.83	0.01
O	0.79	0.04	0.95	0.00
P	0.78	0.02	0.53	0.01

Table 4.8
Hierarchic relocation clustering

Euclidean distance gives reasonable levels of predictive performance across the range of data sets and, as noted earlier, gives results that are comparable with those obtained from several other similarity measures (Willett, 1984b). The results obtained with the size difference measure are rather different in that they tend to be either very much better than those obtained with the Euclidean distance and with the other methods, e.g., the molar refractivity, tadpole narcosis, local anaesthetic and ether toxicity data, or very much worse than them, e.g., the quinazolines, penicillins and heats of vaporisation. It was noted above that this measure is very different from the others in that it does not take account of the actual substructures used to characterise structures and clusters of structures. The measure would hence be expected to give low levels of predictive performance for those data sets where the property under consideration is strongly dependent upon the functionalities that are present. This is what happens in practice since heats of vaporisation are highly dependent upon functionality, while the biological activities of the penicillin and quinazoline data sets are dependent upon logP, and hence upon functionality. In the cases where the size difference measure gives noticeably good results, the tadpole narcosis and local anaesthetic data sets are very diverse in both size and complexity; molar refractivity is dependent upon functionality to some extent, but this might be out-weighed for these compounds by sheer molecular size, i.e., the numbers of fragments present in a structure, while the ether toxicity data set, where all of the molecules share a common functionality, might also be strongly dependent upon molecular complexity. While the size difference measure is unlikely to be sufficiently discriminating for use with large numbers of molecules, it is clear that it may well be appropriate for the analysis of relatively small numbers of compounds where molecular bulk is thought to be of importance in determining molecular activity.

It will be clear that the use of hierarchic relocation clustering has resulted in a generally high level of predictive performance, despite the fact that the methods are much less computationally demanding than the more complex hierarchic procedures considered in previous sections. It was accordingly decided to study a wide range of inexpensive clustering methods, specifically non-hierarchic methods, in more detail. The results of these studies are considered in the next section.

4.3.5 Non-hierarchic clustering methods

The non-hierarchic tests involve several different types

of non-hierarchic clustering method. The first procedure is a simple leader algorithm, as described in Chapter 2. Initial tests with this method showed that the effectiveness of the classification is crucially dependent upon the threshold similarity used. In particular, high values often result in large numbers of singletons, i.e., clusters containing only a single compound, and long run-times owing to the number of molecule-centroid similarity calculations required. It was decided to choose for each data set a threshold which gives a number of final clusters approximately equal to N/4; singleton clusters are then eliminated by assigning each such molecule to that cluster with the centroid most similar to the fragment list describing the molecule. This assignment procedure was adopted to eliminate singletons resulting from all of the different clustering methods described here.

A theoretical basis for a single pass clustering method has been described by Can and Ozkarahan (1984). Although rather more complicated than the leader algorithm, this method provides a rational approach to the selection of the cluster centres, and also eliminates the order dependence inherent in most single pass methods. In essence, the method involves an analysis of the data to identify those molecules which have the greatest numbers of discriminating fragments in common with other compounds. Such molecules can then be used to select the desired number of cluster centroids, N/4 in this case, by selecting the first N/4 compounds after ranking to act as the cluster centres. It will thus be realised that the Can-Ozkarahan clustering method involves the identification of the initial clusters prior to clustering, as with relocation clustering, but also involves only a single pass through the data set.

Two types of relocation method are tested. The first of these uses the clusters resulting from the leader algorithm to give the centroids which form the basis for the subsequent relocation. Alternatively, the centroids for the relocation are derived using the procedure described by Hartigan and Wong (1979). In this, the centroid of the entire data set is calculated, and then the individual structures ranked in order of decreasing similarity from this centroid. A set of N/4 clusters is then obtained by one of the two following procedures:

* the first 4 compounds in the ranked list are allocated to the first cluster, the second 4 compounds to the second cluster, etc.
* the first N/4 compounds are used as the cluster seeds.

These two procedures will be referred to as RA and RB respectively.

Finally, two means are available for implementing the Jarvis-Patrick method. The first is the simple procedure described in Section 2.3 where a pair of compounds which occur in each other's NNTABLE must share at least M near neighbours in common if they are to be grouped into the same cluster. This <u>unweighted</u> approach assumes that all of the near neighbours are of equal importance. Alternatively, a <u>weighted</u> approach takes account of the relative orderings of the neighbours in the two near neighbour lists. If some molecule occurs at the P'th position in one list and the Q'th position in the other, and each such list contains NUM near neighbours, then a contribution of (NUM+1-P)*(NUM+1-Q)is made to M, the overall degree of similarity between I and J. Thus if NUM=20 and if the same compound occurs at the first and fifth positions in the near neighbour lists of I and J, a contribution of 320 is made to M. For both variants of this method a range of different thresholds was tested, and those selected which give a final number of clusters approximately equal to N/4 so as to allow comparison with the other methods that are tested.

The first results are those obtained from the use of the leader algorithm, and from the relocation method with an initial partition deriving from the leader algorithm. In both cases, a total of five classifications was generated and the correlation coefficients listed are mean values averaged over the five runs; also listed are the corresponding standard deviations(Table 4.9).

Data	L		LR	
A	0.85	0.26	0.85	0.26
B	0.82	0.05	0.81	0.04
C	0.79	0.05	0.79	0.05
D	0.62	0.16	0.63	0.23
E	0.74	0.04	0.72	0.08
F	0.94	0.01	0.94	0.01
G	0.84	0.08	0.84	0.06
H	0.93	0.03	0.93	0.03
J	0.85	0.05	0.85	0.04
L	0.75	0.09	0.39	0.02
M	0.70	0.07	0.72	0.06
N	0.77	0.04	0.79	0.07
O	0.77	0.16	0.81	0.13
P	0.88	0.03	0.89	0.02

Table 4.9

Property prediction using the leader (L) and leader/relocation (LR) methods

These results show a high level of predictive ability which is comparable with the results from any of the previous methods that have been discussed. However, many of the standard deviations are very large, much more so than in the case of the relocation methods tested in the previous section, thus implying that the utilities of the final classifications are strongly dependent upon the order in which the set of compounds is processed. Such a dependence is quite unsatisfactory for any practical clustering system since it implies that a superior classification might be obtained simply by a re-ordering and re-clustering of the data.

The correlation coefficients for the Can-Ozkarahan, relocation and Jarvis-Patrick experiments are shown below in Table 4.10. All of the methods produce classifications which are invariant under a re-ordering of the data set, and thus only a single coefficient is reported.

The Can-Ozkarahan method consistently gives results inferior to those obtained with the other methods, and in some cases the difference is very marked, e.g., the amino acid, penicillin and tadpole narcosis data sets. The two relocation procedures give rather better results, but the most consistent level of predictive performance is provided by the Jarvis-Patrick classifications, with the weighted method doing slightly better than the simpler, unweighted approach. Further experiments are reported by Willett et al. (1986) who present results obtained with the Tanimoto coefficient, rather than with the Euclidean distance as used here; these results are very similar to those shown above and confirm that the Jarvis-Patrick method is the most appropriate method of those tested. Moreover, the results are comparable with, or superior to, those given by the hierarchic agglomerative methods, where the computational requirements are very much greater.

Data	CO	RA	RB	JPU	JPW
A	0.39	0.68	0.68	0.90	0.96
B	0.72	0.85	0.68	0.85	0.86
C	0.54	0.60	0.61	0.73	0.74
D	0.53	0.64	0.70	0.85	0.81
E	0.67	0.70	0.69	0.80	0.81
F	0.93	0.94	0.90	0.94	0.94
G	0.35	0.84	0.85	0.81	0.86
H	0.90	0.94	0.90	0.94	0.94
J	0.82	0.82	0.68	0.88	0.89
L	0.75	0.61	0.71	0.72	0.75
M	*	0.51	0.64	0.67	0.79
N	0.73	0.72	0.75	0.79	0.83
O	0.28	0.93	0.64	0.84	0.85
P	0.73	0.82	0.72	0.84	0.86

Table 4.10

Property prediction using the Can-Ozkarahan (CO), two relocation (RA and RB), and unweighted (JPU) and weighted (JPW) Jarvis-Patrick clustering methods

4.4 CLUSTERING OF CHEMICAL STRUCTURE FILES

The comparative tests in the previous section have clearly demonstrated that it is possible to group together compounds with similar property characteristics by applying cluster analysis procedures to simple, fragment-based representations of chemical structures. However, the experiments so far have been restricted to data sets containing only small numbers of structures; this section considers the clustering of large files of molecules typical of those encountered in operational chemical information systems. Two applications are described, these being the selection of compounds for activity testing in biological screens, and the clustering of the outputs resulting from chemical substructure searches. The Jarvis-Patrick method is used for both applications since it is not only highly effective in operation, but can also be implemented so as to allow the clustering of very large files of compounds.

4.4.1 Selection of compounds for activity testing

As discussed in Section 1.4, drug discovery programmes often include empirical, or random, screening of very large numbers of compounds as a means of identifying potential lead compounds. Computer assistance in the selection of compounds for screening has traditionally been provided by the use of substructural analysis techniques which require a fair number of structures of known activity; alternatively, if only a single active compound is available, the ranking methods of Chapter 3 may be applied to identify structurally related compounds which may also exhibit the desired activity. However, such methods cannot be used unless at least some activity data is available. In such cases, a useful facility would be the ability to isolate some small subset of the compounds in a machine-readable file, this subset being carefully chosen to reflect all of the major classes of structure present: the screening programme would then be based upon a systematic testing of each structural class in turn. Specifically, it is suggested that the subset may be obtained by clustering the compounds in the file, and then selecting one molecule from each resulting cluster to represent the compounds in that cluster.

Pfizer Central Research (U.K.) at Sandwich has for some years maintained a Structural Representatives File (SRF), a sub-set of approximately 5% of the total compound collection. The SRF comprises compounds for which sufficient material is available for use in screening, the compounds having been selected so as to be representative of the structural variation present in the total collection. The method of selection has to date been intellectual, and somewhat ad hoc, generally amounting to the inclusion of compounds when they contain previously unrepresented ring systems or functionalities. There are several advantages associated with the use of an SRF generated by purely automatic means. Thus, a complex and time-consuming manual operation involving highly trained staff is replaced by a cheap automatic procedure; an effective clustering procedure should help to ensure that no classes of compounds are overlooked when selecting structures for testing; the existence of a classification can help to dictate which compounds are tested next in a programme, since the identification of one active compound would suggest that the other members of that compound's cluster should also be investigated. Finally, the possibility of altering various parameters within the clustering program allows the creation of different types of SRF to suit different screening requirements. For example, when there is screening capacity available for many compounds to be tested, it should be possible to select molecules from a classification in which there are large numbers of small clusters; conversely, a classification that comprised a smaller number of relatively large clusters would be more appropriate when only a limited amount of screening capacity is available.

The Pfizer stores file is, composed, in large part, of the set of compounds which was used for the similarity searching experiments described in Section 3.3. As noted there, the molecules are characterised by fragment bit strings representing the presence or absence of a range of atom-centred and bond-centred fragments; conversely, most of the property prediction experiments reported above involved frequency of occurrence data and involved just a single fragment type. The Pfizer stores file was clustered using the Jarvis-Patrick method in two stages. In the first, an overnight batch program was used to obtain the 20 nearest neighbours for each of the compounds in the file; these neighbours were identified using the efficient nearest neighbour search algorithm, algorithm F, which is described in Section 5.3.3 and which was used in the browsing search mechanism discussed in Section 3.3.3. The near neighbour table was then written to a disc file which formed the input to the second, and interactive, clustering program. As finally implemented, the user of this program can specify that either a weighted or an unweighted classification should be produced, and can specify the particular similarity thresholds which are to be used, although default values are adopted if required. The file is clustered according to the parameters specified, and the number of clusters generated is output together with the number of singleton clusters. The user of the program is then given the option of clustering these singletons by assigning them to that cluster to which they are most similar. The cluster centroids for this assignment are created by a thresholding procedure so as to obtain centroid bit strings entirely analogous to those which are used to represent each of the individual molecules; specifically, the bit corresponding to some particular screen is set in the centroid bit string if it occurs in at least S/4 of the compounds in a cluster of size S.

As well as providing a means for the elimination of singleton clusters, the centroids may also be used to obtain an ordering of the compounds within a given cluster. This can be done by calculating the similarities between the centroid and the bit strings characterising each of the structures within that cluster; the resulting similarities may then be used to obtain an ordering of the compounds. Whilst the centroid summarises the structural features present within a cluster, it may be useful in some circumstances to select an actual molecule to characterise the set of compounds within some cluster; a natural choice for such a <u>representative</u> is that compound which is most similar to the centroid, i.e., the molecule at the top of the ranking. Such representatives might be required, for example, when selecting one compound from each cluster for activity testing.

Figures 4.4(a) and (b) show two of the clusters which were obtained from an unweighted Jarvis-Patrick classification with a similarity threshold of 3. Further experimental results, using both weighted and unweighted classifications and with a range of similarity thresholds, are presented by Willett et al. (1986). It is clear that the Jarvis-Patrick method provides an effective means of grouping compounds, in that the clusters contain molecules which are highly similar to each other while being quite distinct from the other clusters in the file. Accordingly, the set of clusters should provide an excellent basis for the selection of representative structures, and this clustering method is now used on a regular basis at Pfizer Central Research (UK) to assist in compound selection.

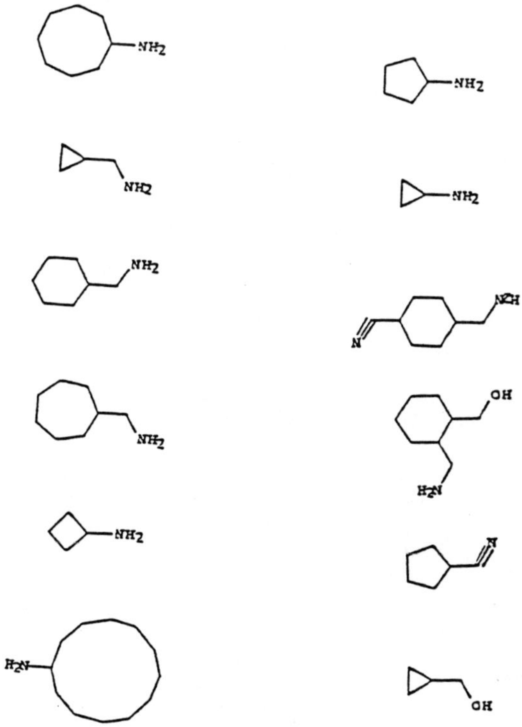

Figure 4.4 (a)

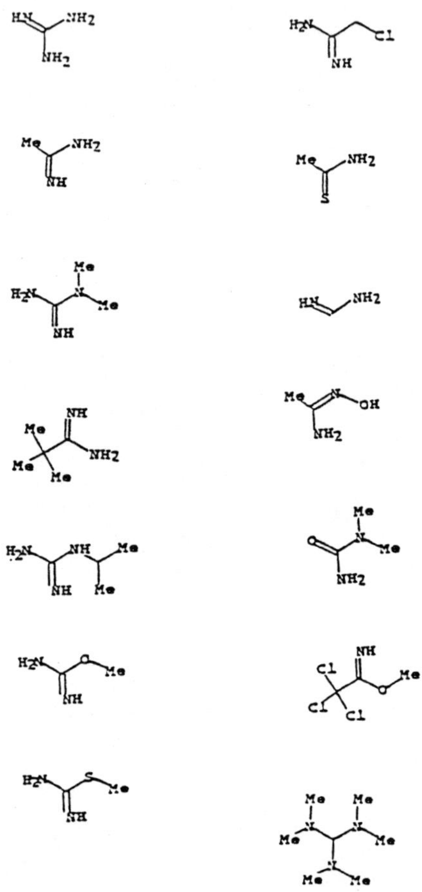

Figure 4.4 (b)

4.4.2 Clustering of substructure search output

As described in Section 1.3, current chemical retrieval systems allow highly efficient and effective substructure searches to be carried out if the user of the system can specify the structural requirements of the query in reasonably specific and precise terms. It is more difficult for such systems to handle generic queries where many alternative atoms or substituent patterns may need to be specified to ensure that all structures of interest are retrieved. In such cases, the recall-oriented nature of the search may result in the retrieval of many compounds which, although satisfying the constraints imposed by the query, are of only marginal relevance to the problem at hand. Section 3.3.2 describes a similarity matching procedure which requires the user to specify a typical compound of interest: this compound is then matched against the molecules retrieved by a substructure search so as to identify those which are most similar to it. In some cases, however, the chemist may not be able to specify such a target compound, or may wish instead to gain an over-view of the main structural classes which are present in the search output.

One way of handling such queries would be to cluster the output of a conventional substructure search into a series of groups of compounds. The resulting classification could then be inspected by selecting a representative compound from each cluster, and displaying the set of such representatives on the terminal. All of the compounds in a particular cluster might then need to be inspected by the chemist if the representative molecule proved to be of interest, while such compounds could be rejected if the representative appeared to be inappropriate.

An operational retrieval system based upon such ideas is described by Willett et al. (1986) for the clustering of searches of the Pfizer stores file used in the previous experiments. In this system, a substructure search is carried out in the normal manner. The bit strings of the compounds retrieved by the search are then passed on for clustering by the Jarvis-Patrick method: thus, for each molecule retrieved in the search, the 20 nearest neighbours within this sub-set are identified, and these nearest neighbour lists then processed as described previously to generate a weighted Jarvis-Patrick classification. The set of cluster representatives is then generated and displayed as the output from the search. As an example, consider a search for optionally substituted chloro-benzenes which retrieves a total of 690 molecules from the stores file. These compounds were clustered to give a total of 55 non-singleton clusters. The five structures that are most similar to the cluster centroids for three randomly selected clusters are shown in Figure 4.5 , where it will be seen that cluster 4 contains carboxyl-substituted mono-benzenes, while the molecules in cluster-26 all contain a conjugated carbonyl linking together 2 benzene rings, one of which contains the specified chloro-substituent. The molecules in cluster 48 again all contain the unsaturated carbonyl, substituted by a cyanide group, but here one of the rings is a saturated N-heterocycle.

Cluster 4

Cluster 26

Cluster 48

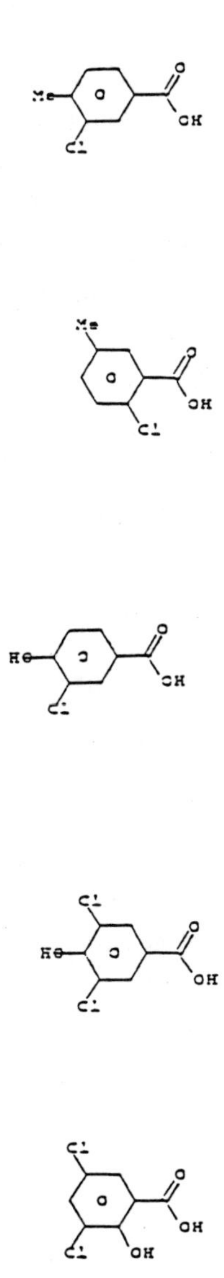

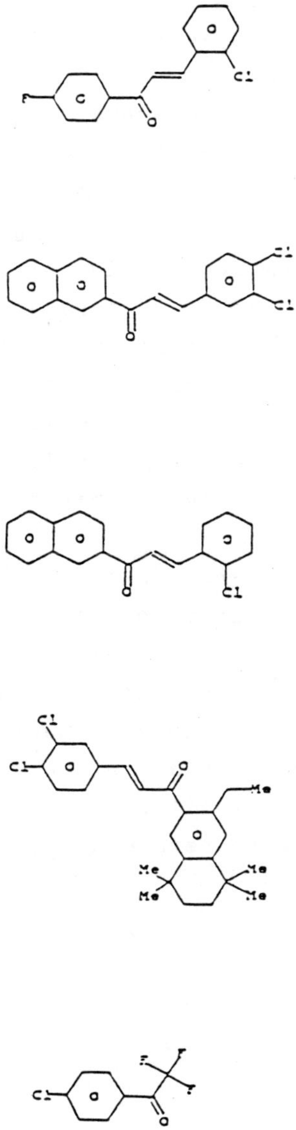

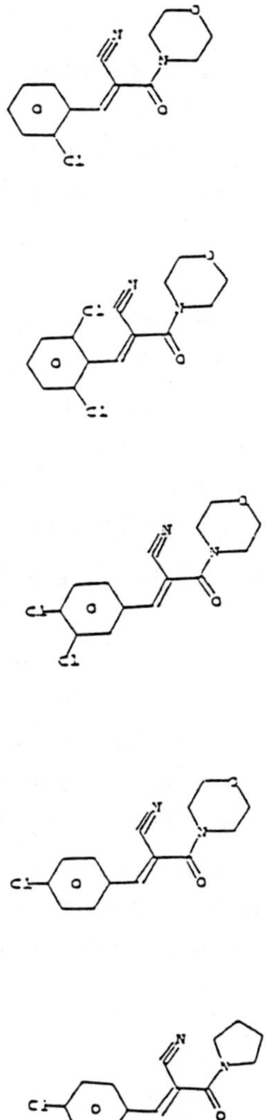

Figure 4.5

A further example involves the retrieval of all bromine-containing molecules. Such a query is, of course, much less specific than would generally be the case in an operational chemical substructure search system; however, such highly generic queries provide a 'worst case' test of system effectiveness. This search retrieves a total of 591 molecules, these being grouped into 47 non-singleton clusters. The first five compounds from four of these clusters, numbers 8, 23, 38 and 39, are illustrated in Figures 4.6(a) and (b), where it will be seen that the clustering procedure results again in the identification of groups of compounds which are relatively homogeneous in nature, but markedly different from the other clusters present. Cluster 8 consists entirely of bromo-alkenes while cluster 23 contains mono- and di-bromo-benzenes; all of the structures in cluster 38 contain a common moiety while the compounds of cluster 39 are characterised by the -CH=NH- linkage between a pair of benzene rings.

The clusters shown here suggest strongly that the classification procedure provides an effective means of summarising the range of structural classes present in the output from a chemical substructure search. When coupled with the similarity-based retrieval mechanisms of Section 3.3, such procedures provide a much wider and more flexible range of search facilities than are provided in current chemical structure retrieval systems.

Cluster 8 Cluster 23

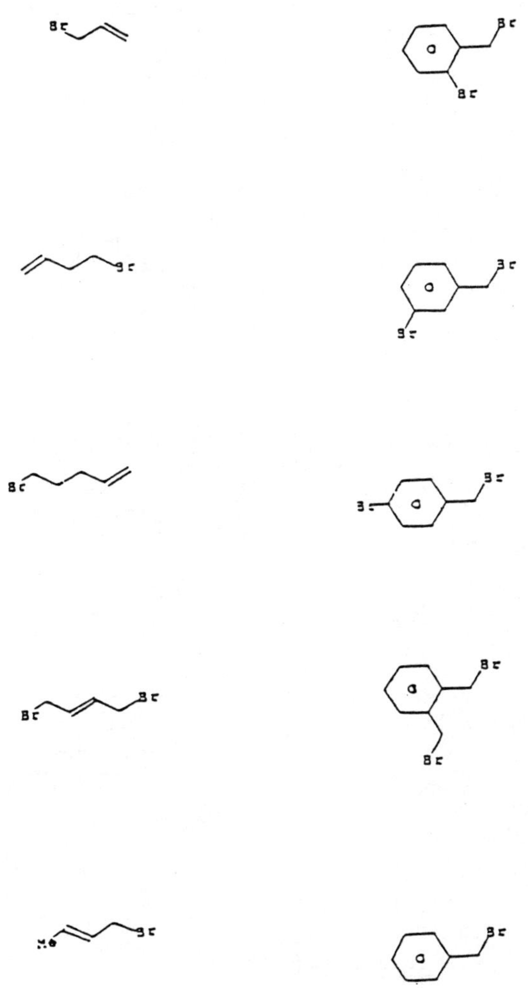

<u>Figure 4.6</u> (a)

Cluster 38

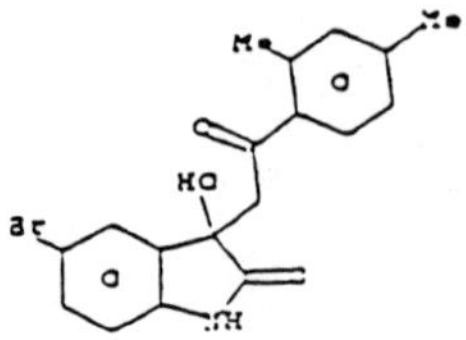

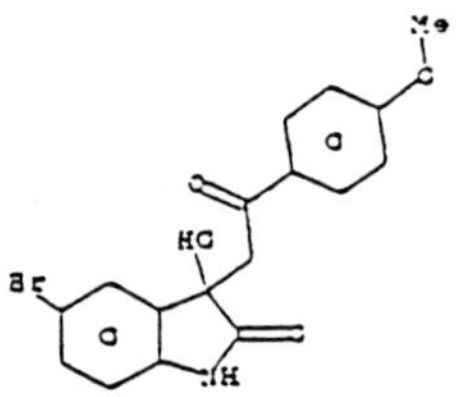

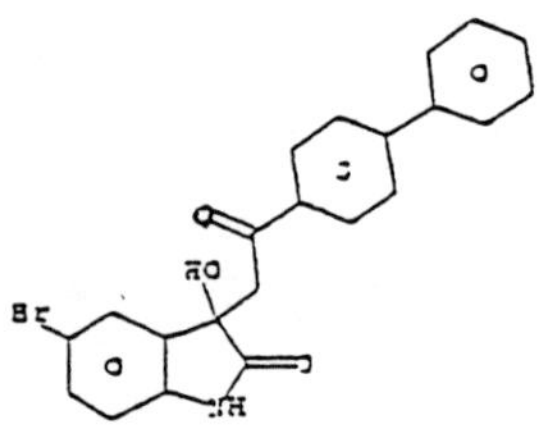

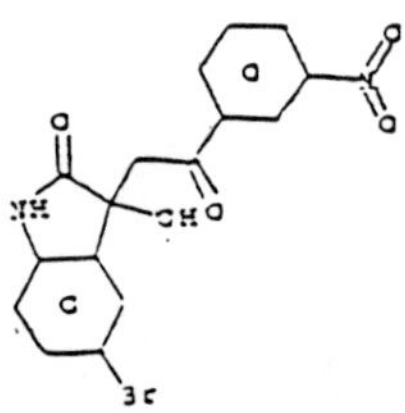

Cluster 39

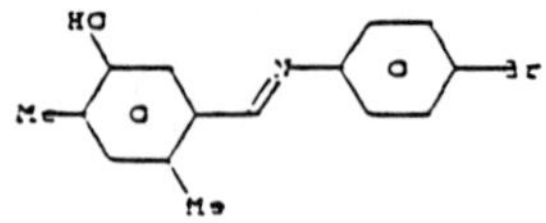

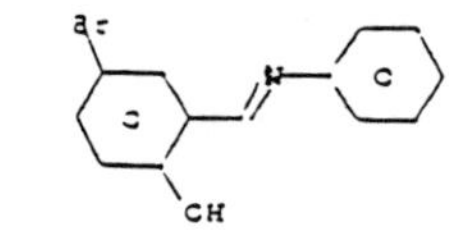

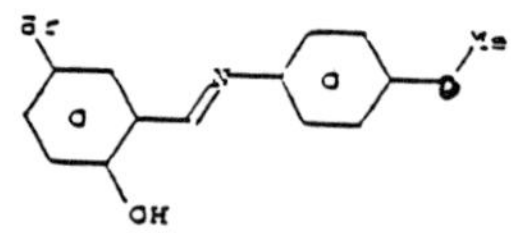

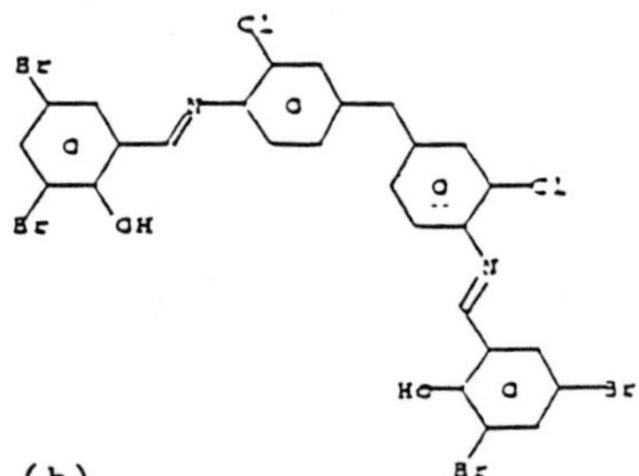

<u>Figure 4.6</u> (b)

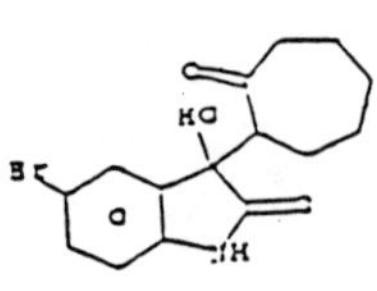

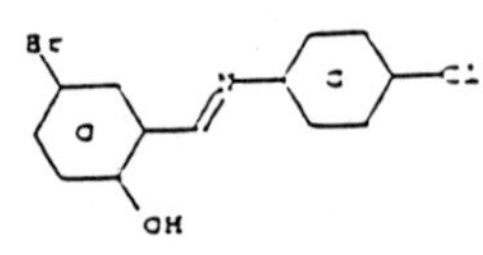

REFERENCES

Adamson, G.W. and Bawden, D. (1981). Comparison of hierarchical cluster analysis techniques for the automatic classification of chemical structures. J. Chem. Inform. Comput. Sci. 21, 204-209.

Adamson, G.W. and Bush, J.A. (1973). A method for the automatic classification of chemical structures. Inform. Stor. Retr. 9, 561-568.

Adamson, G.W. and Bush, J.A. (1975). A comparison of the performance of some similarity and dissimilarity measures in the automatic classification of chemical structures. J. Chem. Inform. Comput. Sci. 15, 55-58.

Bailey, T. A. and Dubes, R. (1982). Cluster validity profiles. Pattern Recognition 15, 61-83.

Bush, J.A. (1977). Automatic Classification and Chemical Structure-Activity Correlation. PhD thesis, University of Sheffield.

Can, F. and Ozkarahan, E.A. (1984). Two partitioning type clustering algorithms. J. Amer. Soc. Inform. Sci. 35, 268-276.

Chu, K. (1974). Applications of artificial intelligence to chemistry. Use of pattern recognition and cluster analysis to determine the pharmacological activity of some organic compounds. Anal. Chem. 46, 1181-1187.

Clifford, H.T. and Stephenson, W. (1975). An Introduction to Numerical Classification. Academic Press, New York.

Colman, G. (1968). The application of computers to the classification of streptococci. J. Gen. Microbiol. 50, 149-158.

Cormack, R.M. (1971). A review of classification. J. Royal. Stat. Soc. Ser. A 134, 321-353.

Dattola, R.T. (1971). Experiments with a fast algorithm for automatic classification. In Salton, G. (ed.) The SMART Retrieval System. Prentice-Hall, New Jersey.

De Clercq, H., Despontin, M., Kaufman, L. and Massart, D.L. (1976). The selection of representative substances by operations research techniques. J. Chromatog. 122, 535-551.

Delaney, M. F. (1981). Modified heuristic for generating tree-structured spectral libraries. Anal. Chem. 53, 2354-2355.

Dubes, R. and Jain, A.K. (1979). Validity studies in clustering methodologies. Patt. Recognit. 11, 235-254.

Dunn, W.J., Greenberg, M. and Callejas, S. (1976). Use of cluster analysis in the development of structure-activity relations for antitumour triazenes. J. Med. Chem. 19, 1299-1301.

Faris, J.S. (1969). On the cophenetic correlation coefficient. Systemat. Zool. 18, 279-285.

Faris, J.S. (1973). On comparing the shapes of taxonomic trees. Systemat. Zool. 22, 50-54.

Fillenbaum, S. and Rapoport, A. (1971). Structures in the Subjective Lexicon. Academic Press, New York.

Fowlkes, E.B. and Mallow, C.L. (1983). A method for comparing two hierarchical clusterings. J. Amer. Stat. Assoc. 78, 553-569.

Griffiths, A., Robinson, L.A. and Willett, P. (1984). Hierarchic agglomerative clustering methods for automatic document classification. J. Docum. 40, 175-205.

Griffiths, A., Luckhurst, H.C. and Willett, P. (1986). Using inter-document similarity information in document retrieval systems. J. Amer. Soc. Inform. Sci. 37, 3-11.

Hansch, C., Unger, S.H. and Forsythe, A.B. (1973). Strategy in drug design. Cluster analysis as an aid in the selection of substituents. J. Med. Chem. 16, 1217-1222.

Harrison, P.J. (1968). A method of cluster analysis and some applications. Appl. Stat. 17, 226-236.

Hartigan, J.A. and Wong, M.A. (1979). A k-means clustering algorithm. Appl. Stat. 28, 100-108.
Hubert, L. (1974). Some applications of graph theory to clustering. Psychomet. 39, 282-309.
Jardine, N. and Sibson, R. (1971). Numerical Taxonomy. Wiley, New York.
Kaufman, L., Pierreux, A., Rousseeuw, P., Derde, M.P., Detaevernier, M.R., Massart, D.L. and Platbrood, G. (1983). Clustering on a microcomputer with an application to the classification of coals. Anal. Chim. Acta 153, 257-260.
Lewi, P.J. (1976). Spectral mapping, a technique for classifying biological activity profiles of chemical compounds. Drug Res. 26, 1295-1300.
Ling, R.F. (1973) Probability theory of cluster analysis. J. Amer. Stat. Assoc. 68, 159-164.
Ling, R.F. (1975). An exact probability distribution on the connectivity of random graphs. J. Math. Psychol. 12, 90-98.
Ling, R.F. and Killough, G.G. (1976). Probability tables for cluster analysis based on a theory of random graphs. J. Amer. Stat. Assoc. 71, 293-300.
MacNaughton-Smith, P., Williams, W.T., Dale, M.B. and Mockett, L.G. (1964). Dissimilarity analysis: a new technique of hierarchical sub-division. Nature 202, 1034-1035.
Massart, D.L., Kaufman, L. and Coomans, D. (1980). An operational research model for pattern recognition. Anal. Chim. Acta 122, 347-355.
Massart, D.L., Kaufman, L. and Esbensen, K.H. (1982). Hierarchical nonhierarchical clustering strategy and application to classification of iron meteorites according to their trace element patterns. Anal. Chem. 54, 911-917.
Massart, D.L. and Kaufman, L. (1983). The Interpretation

of Analytical Chemical Data by the Use of Cluster Analysis. Wiley, New York.

Massart, D.L., Plastria, F. and Kaufman, L. (1983). Non-hierarchical clustering with MASLOC. Patt. Recognit. 16, 507-516.

Miyashita, Y., Takahashi, Y., Yotsui, Y., Abe, H. and Sasaki, S.I. (1981). Application of pattern recognition to structure-activity problems. Use of minimal spanning tree. Anal. Chim. Acta 133, 615-620.

Murtagh, F. (1983). A New Approach to the Comparison of Hierarchic Clustering. University College, Dublin.

Murtagh, F. (1984). Structure of hierarchic clusterings: implications for information retrieval and multivariate data analysis. Inform. Proc. Manag. 20, 611-617.

Phipps, J.B. (1971). Dendrogram topology. Systemat. Zool. 20, 306-308.

Ritter, G.L. and Isenhour, T.L. (1977). Minimal spanning tree clustering of gas chromatographic liquid phases. Comput. Chem. 1, 145-153.

Rohlf, F.J. (1974). Methods of comparing classifications. Ann. Rev. Ecol. and Systemat. 5, 101-113.

Rubin, V. and Willett, P. (1983). A comparison of some hierarchal monothetic divisive clustering algorithms for structure property correlation. Anal. Chim. Acta 151, 161-166.

Saggers, D.T. (1974). The application of the computer to a pesticide screening programme. Pest. Sci. 5, 341-352.

Saunders, R. and Funk, G.M. (1977). Poisson limits for a clustering model of Strauss. J. Appl. Prob. 14, 776-784.

Schultz, J. and Hubert, L. (1973). Data analysis and connectivity of random graphs. J. Math. Psychol. 10, 421-428.

Sibson, R. (1973). SLINK: an optimally efficient algorithm for the single-link cluster method. Comput. J. 16, 30-34.

Sneath, P.H.A. (1966). Relations between chemical structure and biological activity in peptides. J. Theor. Biol. 12, 157-195.

Sokal, R.R. and Rohlf, F.J. (1962). The comparison of dendrograms by objective means. Taxon. 11, 33-40.

Strauss, D.J. (1975). Model for clustering. Biometrika 62, 467-475.

Takahashi, Y., Miyashita, Y., Abe, H. and Sasaki, S.I. (1980). A structure-biological activity study based on cluster analysis and the non-linear mapping method of pattern recognition. Anal. Chim. Acta 122, 241-247.

Vorhees, E. (1985). The Efficiency and Effectiveness of Hierarchic Agglomerative Clustering in Document Retrieval. PhD thesis, Cornell University.

Waterman, M.S. and Smith, T.F. (1978). On the similarity of dendrograms. J. Theor. Biol. 73, 789-800.

White, R.F. and Lewinson, T.M. (1977). Probabilistic clustering for attributes of mixed type with biopharmaceutical applications. J. Amer. Stat. Assoc. 72, 271-277.

Willett, P. (1982). A comparison of some hierarchal agglomerative clustering algorithms for structure property correlation. Anal. Chim. Acta 136, 29-37.

Willett, P. (1984a). A note on the use of nearest neighbours for implementing single linkage document classifications. J. Amer. Soc. Inform. Sci. 35, 149-152.

Willett, P. (1984b). Evaluation of relocation clustering algorithms for the automatic classification of chemical structures. J. Chem. Inform. Comput. Sci. 24, 29-33.

Willett, P. (1985). The validity of chemical classifications. J. Chem. Inform. Comput. Sci. 25, 78-80.

Willett, P., Winterman, V. and Bawden D. (1986). Implementation of non-hierarchic cluster analysis methods in chemical information systems. Selection of compounds for biological testing and clustering of substructure search output. J. Chem. Inform. Comput. Sci. In press.

Williams, W.T. and Clifford, H.T. (1971). On the comparison of two classifications of the same set of elements. Taxon. 20, 519-522.

Williamson, R.E. (1974). Real-Time Document Retrieval. PhD thesis, Cornell University.

Wishart, D. (1978). CLUSTAN 1C User Manual. Program Library Unit, Edinburgh University.

Zahn, C.T. (1971). Graph-theoretical methods for detecting and describing gestalt clusters. IEEE Trans. Comput. C-20, 68-86.

Zhitkov, G.N. (1970). Classification of structural elements in analysing complex images. Automatic Documentation and Mathematical Linguistics 4, 19-25.

Zupan, J. (1980). A new approach to binary tree-based heuristics. Anal. Chim. Acta 122, 337-346.

Zupan, J. (1982). Hierarchical clustering of infrared spectra. Anal. Chim. Acta 139, 143-153.

Zupan, J. (1983). Clustering of Large Data Sets. Research Studies Press, Chichester.

CHAPTER 5
Nearest Neighbour Searching Algorithms

5.1 NEAREST NEIGHBOUR SEARCHING

5.1.1 Introduction

The two previous chapters have discussed the use of similarity searching and clustering in chemical structure handling systems, but without any detailed description of how such operations can be carried out in a manner that minimises the use of computer memory and processor resources. Efficient algorithms to carry out these tasks form the subject of this chapter. Much of the material presented here derives from the work that has been done over the last few years in the design of best match search algorithms for document retrieval systems. In such systems, queries and documents are described in the form of a list of weighted or unweighted keyword descriptors, and the query is searched against a file of documents to identify those that are most similar to it. Such systems are closely related to the chemical structure environment since, in both cases, the data files contain records, each of which is characterised by a small number of descriptors selected from a much larger number of possible descriptor types.

The obvious, brute-force algorithm for best match searching is to compute the distance between the query, hereafter denoted by Q, and each of the N compounds in the file, and then to select as the nearest neighbour that molecule which is closest to the query. This algorithm involves no less than N distance calculations and is thus much too time-consuming for all but the smallest files. To overcome this problem, methods have been suggested that avoid the calculation of most of the distances while still calculating the distances for those few records which are, in fact, near to the query. Several types of criterion have been suggested to reduce the number of calculations required, including the projection of the multi-dimensional records onto a space of lower dimensionality where most of the distance calculations are performed (Friedman et al., 1975; Lee et al., 1976), and the grouping of records into clusters so that several records may be searched, or eliminated from the search, simultaneously (Burkhard and Keller, 1973; Yuval, 1976; Friedman et al., 1977; Shapiro, 1977; Bentley et al., 1980). A recent review of these studies is given by Murtagh (1984b).

5.1.2 Nearest neighbour searching in chemical files

Unfortunately, most of the algorithms described in the previous section are not directly applicable to nearest neighbour searching in the context of chemical structure files for two reasons.

Firstly, the algorithms generally assume that the attributes are continuous variables, whereas the chemical structures discussed in this book are normally represented by fragment descriptors that can take only integer values. It is, of course, possible for some real-valued weighting scheme to be used but, as has been noted in Chapter 3, this may not be of any noticeable use in practice. Moreover, in most current structure search systems, such as those described in Chapter 1, the primary molecular representation is in binary terms, denoting merely the presence or absence of a particular fragment substructure.

Secondly, it is usually assumed that M, the dimensionality of the space that contains the records, is very low, typically 2 or 3, so that multiplicative terms involving M in the equation describing the complexity of the search algorithm may be neglected. In a chemical structure search system, conversely, many hundreds or thousands of substructures may be used to characterise the compounds in the file, and such algorithms are thus quite impracticable. An example of this is the O(log N) procedure of Friedman et al. (1977), which Weiss (1981) suggests involves a constant of proportionality of about 1.6^M. Again, the constant expected time search algorithm of Bentley et al. (1980) involves the inspection of all of the 3^M-1 cells adjacent to a given cell in the space, an undertaking that is quite infeasible if the dimensionality is at all large; it also assumes that the data records have been chosen independently from a uniform point distribution throughout the space. As a further example of such problems, Kittler (1978) describes a heuristic for nearest neighbour searching using the city block distance, the efficiency of which is proportional to $M^{-1/2}$.

5.1.3 Formulation of the problem

Despite these problems, a few workers have reported search procedures that are suitable for chemical structure applications. Naegeli and Clerc (1974), Van Marlen and Van den Hende (1979) and Rasmussen et al. (1979) describe best match retrieval algorithms for use with binary-encoded files of mass spectra, while Weiss (1981), Smeaton and Van Rijsbergen (1981), Eastman and Weiss (1982), Murtagh (1982) and Perry and Willett (1983) have studied best match searching in document retrieval systems. This chapter describes the application of some of this work to the calculation of inter-molecular similarities in chemical structure files. The main emphasis is on the identification and retrieval of that single compound which is closest, or most similar, to a given query structure. However, the procedures may be generalized to a range of other closest point problems, e.g., the identification of the 20 nearest neighbours, and to the matching of molecules with clusters in both hierarchic and non-hierarchic clustering methods.

The nearest neighbour search algorithms that are discussed fall naturally into two main groups, arising from the two principal file structures that are used in current chemical retrieval systems. The next section describes algorithms that have been designed for use with serial files, while Section 5.3 describes those that are intended for use with inverted file systems. Empirical studies of the performance of these algorithms are reported in Section 5.4, and the chapter concludes with some comments on the practical implementation of the algorithms.

The algorithms are presented in a PASCAL-like notation. It is assumed that the compounds in a file of N structures are stored for search as a bit map, B, in which bit B[I,J] is set to one if the J'th fragment screen has been assigned to the I'th molecule, $1 <= I <= N$, and $1 <= J <= M$, where M is the number of different fragment types. This bit map can be regarded either as a serial file, in which the bit strings are inspected in sequence one after the other, or as an inverted file, in which access is available to all of the compounds possessing a specific fragment screen. The query structure, Q, that is to be searched against the N compounds is assumed to be represented by a comparable M-element bit string. For simplicity, the number of non-zero bits in this string will be denoted by QSIZE and the number of fragments in common with the I'th molecule by COMMON; further, the number of non-zero bits in the I'th molecular bit string will be denoted by SIZE[I]. In addition to the bit string Q, it is assumed that an integer array, QA, is available in which the I'th element, 1 <= I <= QSIZE, contains the index for the I'th non-zero element of Q; this array will help to clarify the presentation of the inverted file algorithms described in Section 5.3.

Two similarity coefficients will be used in the presentation and evaluation of the algorithms. The simpler of these is COMMON, while normalised coefficients will be exemplified by the Dice coefficient 2*COMMON/(QSIZE+SIZE[I]).

5.2 NEAREST NEIGHBOUR SEARCHING USING SERIAL FILES

5.2.1 A simple algorithm

An obvious, and very inefficient, algorithm for best match searching is shown below in Algorithm A. Here, B[I,] corresponds to the I'th row of the bit map B, and thus represents the bit string characterising the I'th molecule. MATCH is a procedure that takes two bit strings, Q and B[I,] in this case, and returns a value, COEFF, which corresponds to the similarity between the two bit strings using the chosen similarity measure. NN contains the identifier for the current nearest neighbour at any point in the search.

```
PROC NNUPDATE (INTEGER I) :
  BEGIN
    MATCH(Q,B[I,]) ;
    IF COEFF > BESTCOEFF THEN
      BEGIN
        BESTCOEFF := COEFF ;
        NN := I
      END
  END ;
BESTCOEFF := 0.0 ;
FOR I := 1 TO N DO NNUPDATE(I).
```

It will be seen that the inefficiency of this search procedure arises from two main factors; first, that each of the N compounds must be matched against Q, and, second, that Q must be inspected for each such comparison. The algorithms presented below attempt to overcome these two problems by a range of strategies.

5.2.2 An upper bound algorithm

A simple heuristic is available for increasing the

efficiency of best match searching when the Hamming distance is used. The file of compounds is partitioned into a series of mutually exclusive groups, in each of which all of the compounds have the same numbers of fragments assigned, i.e., equal values for SIZE[I]. If the G'th such group contains compounds with a size value of GSIZE[G], the groups are searched in increasing order of the value of |QSIZE-GSIZE[G]|. This corresponds to the minimum possible number of bit mis-matches between Q and a compound with GSIZE[G] fragments assigned to it, and thus provides a simple means of establishing a lower bound to the Hamming distance. Van Marlen and Van den Hende (1979) found that the use of this heuristic substantially increased the efficiency of best match retrieval in an experimental mass spectral search system. The approach is closely related to the reference point algorithms described by Burkhard and Keller (1973) and by Shapiro (1977), who describe low dimensionality nearest neighbour searching with similarity measures that obey the triangle inequality; a modified form of the Burkhard-Keller algorithm is described by Nevalainen and Katajainen (1982). Accordingly, the analysis presented in their papers is not applicable to searches that involve non-metric measures such as the similarity coefficients discussed in Chapter 3: the basic procedure can, however, be modified so that it may be used with such measures.

For a group of compounds, each containing GSIZE[G] fragments, the maximum possible number of fragments in common with Q, MAXPOSS, is given by MIN(QSIZE,GSIZE[G]) where MIN is an operator that returns the smaller of the two subsequent values. From this expression, an upper bound to the value of the Dice coefficient is given by 2*MAXPOSS/(QSIZE+GSIZE[G]). This upper bound value may be calculated for each distinct value of GSIZE[G] that is present in the file, and the structures then matched against Q in decreasing order of the upper bound until the current nearest neighbour has a higher match value than the upper bound for the next block of compounds. This algorithm, algorithm B, is summarised below.

```
FOR G := 1 TO BNUM DO
  IF GSIZE[G] > QSIZE THEN
    UPB[G] := 2*QSIZE/(QSIZE+GSIZE[G])
  ELSE UPB[G] := 2*GSIZE[G]/(QSIZE+GSIZE[G]) ;
SORT(UPB) ;
BESTCOEFF := 0.0 ;
FOR G := 1 TO BNUM WHILE BESTCOEFF > UPB[G] DO
  FOR each compound, I, in this group DO NNUPDATE(I).
```

In the pseudo-code above, it has been assumed that there is a total of BNUM distinct values of GSIZE present in the file, and that the value for the G'th such group of compounds is GSIZE[G]; SORT(UPB) is a procedure that sorts the groups of compounds into decreasing order of the upper bound similarity, UPB. It may be noted that most of the similarity coefficients in common use are symmetric, i.e., SIM[A,B]=SIM[B,A], and this may be taken into account since the upper bound ordering will be such that structures are inspected in the order of size QSIZE, QSIZE+1 or QSIZE-1, QSIZE+2 or QSIZE-2 etc.

5.2.3 Use of representatives

The efficiency of the search may be increased further if a mechanism is available for the grouping of individual compounds, having the same SIZE values, into blocks, each of which is characterised by a bit string obtained from the logical union of all of the molecules in that block: the resultant bit string will be referred to here as a signature. Such ideas have been discussed by several workers (Burke and Rickman, 1973; Vallarino, 1976; Pfalz et al., 1980) as a means of reducing disc accesses in partial match search systems, but not as a means of reducing the number of comparisons in best match searching. The query structure is matched initially against the signature for the G'th block to identify the number of common fragments, RCOMMON; given this information, an upper bound to the number of fragments in common between Q and any of the molecules in that block, MAXPOSS, is given by MIN(RCOMMON,GSIZE[G]) where GSIZE[G] is the SIZE value of the compounds in the block. An upper bound may then be calculated to the value for the Dice coefficient between Q and any of the structures in this block using 2*MAXPOSS/(QSIZE+GSIZE[G]). The compounds within the block need not be considered for matching individually against Q if the calculated upper bound value is less than the observed match value for a previously inspected structure.

As the signature for a block is the union of all of the individual bit strings representing compounds within the block, the number of bits set in the signature will depend in large part upon the number of compounds in the block and thus, since the size of the collection is a constant, upon the number of blocks, NBLOCKS. When NBLOCKS is small, there will be many non-zero bits in each signature and thus the upper bound values that are calculated for each block are likely to be high, which will result in many of the blocks needing to be inspected in detail. Conversely, a high value for NBLOCKS implies that each individual block will be small, and thus that the number of non-zero bits in each signature will also be quite low; accordingly, the upper bound that is calculated is unlikely to be grossly different from the actual similarity values obtained if the compounds comprising that block need to be matched against Q. Thus, for a given set of structures and queries, it is to be expected that there will be some value for NBLOCKS that will minimise the total number of similarity calculations that has to be made.

5.3 NEAREST NEIGHBOUR SEARCHING USING INVERTED FILES

5.3.1 Introduction

The second group of search algorithms to be discussed here is based upon the inverted file organisation. The inverted file to a collection of compounds contains a set of lists, each of which contains pointers to those molecules that contain one particular member of the set of fragment substructures that has been used to characterise the molecules in the file. The importance of the inverted file for best match searching is that it provides a rapid means for the identification of those structures which share fragments in common with a query molecule; in the discussion that follows, the inverted file lists corresponding to the fragments in a query structure or substructure will be referred to as query lists. Thus, the identifiers of compounds containing query fragments will appear in one or more of the QSIZE query lists. Conversely, compounds whose identifiers do not appear in any of these lists do not contain any of the query fragments; such molecules must have zero-valued similarities with the query, and thus need not be considered as potential nearest neighbours. It should be noted that this assumes that a similarity coefficient, such as the Dice or cosine coefficient, is being used to determine the degree of similarity between Q and each of the structures in the file: a discussion of other types of coefficient is given in Section 5.5.

A simple, inverted file, best match search involves the processing of all of the molecules pointed to by each of the query lists. If the current compound at some point in the search has not already been matched against the query, the bit string of that molecule is retrieved and its similarity with Q determined: if the resulting similarity value is greater than that for the current best match the structure becomes the current nearest neighbour. This algorithm, C, is summarised below where INSPECTED is a Boolean vector, the I'th element of which is set to TRUE once the I'th structure has been matched against Q. This algorithm gives a basis of performance against which the inverted file algorithms described in the next two sections may be compared.

```
FOR I := 1 TO N DO
  INSPECTED[I] := FALSE ;
BESTCOEFF := 0.0 ;
FOR J := 1 TO QSIZE DO
  FOR I := 1 TO N DO
    IF B[I,QA[J]] AND NOT INSPECTED[I] THEN
      BEGIN
        NNUPDATE(I) ;
        INSPECTED[I] := TRUE
      END.
```

5.3.2 Upper bound algorithms

The two algorithms to be discussed here are based upon a simple, but extremely powerful, heuristic which is as follows. At some point in a best match search using algorithm C, assume that the structures in the first J-1 of the QSIZE query lists have been matched against Q. Then any previously uninspected structure can have, at most, QSIZE-J+1 fragments in common with Q since, if it had more than this, it would have already appeared in one of the first J-1 query lists. This information may then be used as the basis for an upper bound calculation to determine the maximum possible similarity between Q and the molecules that have not yet been compared with it.

The first use of this heuristic was by Smeaton and van Rijsbergen (1981) who proposed a search procedure, algorithm D, which involves two pre-processing steps that are used to control the number of compounds inspected. At the start of a search, the query fragments are sorted into order of increasing collection frequency so that the inverted file lists with the smallest numbers of non-zero elements are processed first. This ensures that if any of the query lists are eliminated from the search, it will be those that correspond to fragments that occur in the largest number of molecules in the file. The second pre-processing step is to note the minimum SIZE value for all molecules which have been assigned a given fragment; this information needs to be calculated just once when the inverted file is created, and can be stored in the directory to the inverted file. Let this minimum value for the J'th query fragment, QA[J], be MINSIZE[QA[J]] and assume that J-1 of the QSIZE query lists have been processed at some point in the search. From the discussion above, MAXPOSS is given by QSIZE-J+1, and thus an upper bound to the Dice coefficient for all previously uninspected compounds in the J'th query list is 2*MAXPOSS/(QSIZE+MINSIZE[QA[J]]). It is possible that MAXPOSS is greater than MINSIZE[QA[J]] and in this case, MINSIZE[QA[J]] in the denominator of the expression is replaced by MAXPOSS, giving 2*MAXPOSS/(QSIZE+MAXPOSS) as the upper bound.

If the upper bound calculated using one of the above two expressions is less than the match value for the current nearest neighbour, BESTCOEFF, none of the uninspected compounds in the J'th query list can possibly be the nearest neighbour, and the list of structure identifiers need not be processed. This algorithm, algorithm D, is presented below.

```
FOR I := 1 TO N DO
  INSPECTED[I] := FALSE ;
BESTCOEFF := 0.0 ;
FOR J := 1 TO QSIZE DO
  BEGIN
    MAXPOSS := QSIZE-J+1 ;
    IF MAXPOSS > MINSIZE[QA[J]] THEN
      UPB := 2*MAXPOSS/(QSIZE+MAXPOSS)
    ELSE UPB := 2*MAXPOSS/(QSIZE+MINSIZE[QA[J]]) ;
    IF UPB > BESTCOEFF THEN
      FOR I := 1 TO N DO
        IF B[I,QA[J]] AND NOT INSPECTED[I] THEN
          BEGIN
            NNUPDATE(I) ;
            INSPECTED[I] := TRUE
          END
  END.
```

Murtagh (1982) has described the use of a further upper bound calculation which is performed for each potential nearest neighbour structure. This calculation is based upon, and in addition to, that described by Smeaton and van Rijsbergen. Specifically, the upper bound is calculated as before, but with MINSIZE[QA[J]] replaced by the SIZE value for the actual structure that is under consideration as the potential nearest neighbour; thus the I'th structure will have an upper bound value of 2*MIN((QSIZE-J+1),SIZE[I])/(QSIZE+SIZE[I])). Since this upper bound is specific to the I'th compound, rather than applying to all of the compounds in the J'th query list, the calculated value is very much lower than that obtained from the use of MINSIZE[QA[J]], and is thus much more likely to be able to eliminate that structure from consideration for matching against Q. The increase in efficiency over the brute-force search is accordingly expected to be very much greater than in the case of algorithm D.

The algorithm, algorithm E, may be described by taking algorithm D, as outlined above, and replacing

```
IF B[I,QA[J]] AND NOT INSPECTED[I] THEN
  BEGIN
    NNUPDATE(I) ;
    INSPECTED[I] := TRUE
  END
```

by the more extended

```
IF B[I,QA[J]] AND NOT INSPECTED[I] THEN
  BEGIN
    IF MAXPOSS > SIZE[I] THEN
      UPB := 2*SIZE[I]/(QSIZE+SIZE[I])
    ELSE UPB := 2*MAXPOSS/(QSIZE+SIZE[I]) ;
    IF UPB > BESTCOEFF THEN
      BEGIN
        NNUPDATE(I) ;
        INSPECTED[I] := TRUE
      END
  END.
```

The upper bound calculations appropriate to other similarity measures are given by Murtagh (1982) who also provides theoretical arguments for believing that the algorithm will perform most efficiently when the Hamming distance is used as the similarity coefficient; this conclusion is not supported by the experimental results of Perry and Willett (1983).

5.3.3 An efficient implementation of algorithm C

A very different inverted file algorithm has been described in the context of document retrieval systems by Matthews and Thomson (1969), Noreault et al. (1977) and by Perry and Willett (1983). This algorithm, algorithm F, again involves the processing of the query lists in sequence. However, the novelty of the algorithm lies in the fact that it is the lists themselves which are used for the calculation of the similarities, whereas the three previous inverted file algorithms merely use the lists to identify those structures which need to be compared with Q.

The algorithm is based upon the addition of the query lists to yield a vector, V. When all of the query lists have been added together, the I'th element, V[I], will contain the number of fragments in common between Q and the I'th structure so that the similarity is given by 2*V[I]/(QSIZE+SIZE[I]). Like algorithm C, this procedure will result in the calculation of all of the non-zero similarities, and it would thus appear to be very much less efficient in operation than the two upper bound searches described in Section 5.3.2. Against this, however, is the fact that determining the number of fragments in common between a pair of bit strings involves scanning both of them, and thus the list of query fragments, QA, needs to be inspected each and every time that a similarity needs to be determined. In algorithm F, conversely, the set of query fragments is inspected once only to identify which inverted file lists are to be added together, and the elimination of the repeated scanning means that the procedure is very fast in operation despite the fact that large numbers of coefficients are evaluated.

A discussion of the use of this algorithm for chemical superstructure searching is given by Willett (1985), while implementation strategies are described by Willett (1981) and by Perry and Willett (1983). The algorithm is as follows:

```
FOR I := 1 TO N DO
  V[I] := 0 ;
BESTCOEFF := 0.0 ;
FOR J := 1 TO QSIZE DO
  FOR I := 1 TO N DO
    IF B[I,QA[J]] THEN V[I] := V[I]+1 ;
FOR I := 1 TO N DO
  BEGIN
    COEFF := 2*V[I]/(QSIZE+SIZE[I]) ;
    IF COEFF > BESTCOEFF THEN
      BEGIN
        BESTCOEFF := COEFF ;
        NN := I
      END
  END.
```

5.4 COMPARATIVE STUDIES OF ALGORITHM EFFICIENCY

5.4.1 Nearest neighbour searching experiments

The efficiencies of the algorithms above have been tested using a set of 2335 structurally disparate compounds from the Index Chemicus Registry System. Each of the molecules is characterised by a bit string describing the augmented atom fragments contained within it, and an additional 100 heterogeneous compounds have been similarly represented to act as queries for which the best matching structure is required.

Algorithm	Simple	Dice
Blocked serial search (B)	0.14	0.09
Smeaton-Van Rijsbergen (D)	0.05	0.11
Murtagh (E)	0.03	0.04

Table 5.1

Table 5.1 lists the mean fraction of the file that needs to be matched against the query structure to identify the nearest neighbour, the average being calculated over the entire set of query molecules. For comparison with the results in this table, the simple serial search, algorithm A, and the efficient inverted file search, algorithm F, give search fractions of 1.0 and 0.78 respectively for both similarity coefficients. In the case of the blocked serial search, the results correspond to the value of NBLOCKS, as defined in Section 5.2, which gives the best overall level of performance. Further results are presented by Willett (1983).

The figures in the table show clearly that the upper bound calculations can substantially reduce the number of compounds that must be matched against a query structure, with algorithm E, in particular, giving a substantial increase in efficiency over the simple algorithm A. However, the overall computational costs will also depend upon other factors such as the pre-processing and additional storage that are needed, and upon the extent to which accesses must be made to backing storage. The pre-processing requirements have been described in previous sections, and discussion will hence be restricted here to the last two factors.

The only additional storage needed for the blocked

serial search, algorithm B, is that required for the signatures since these form an additional search file in themselves. Storage equivalent to NBLOCKS M-bit strings will be required since there is one signature per block, while additional space will be required for an integer table giving the value of GSIZE[G] for each block. The two upper bound searches need to store the MINSIZE values for each fragment type in the inverted file, and also need to provide rapid access to the table containing the SIZE values in the case of algorithm E. As well as the SIZE values, algorithm F requires a further N integers of internal storage for the vector V; however, direct access to the table of SIZE values is not required since all of the coefficients are evaluated at the same time, once all of the query lists have been processed.

It has been assumed so far that the entire bit map can be held within the main store of the computer: this is unlikely to be possible if many thousands of structures need to be searched, and consideration must hence be given to the effect of disc accesses on the search algorithms. In the case of algorithm B, a considerable amount of seek time will be needed to allow the scanning of the file in the order dictated by the calculated upper bound values. The inverted file searches, algorithms C to E, will require accesses for the query lists and, more importantly, accesses for each of the individual bit strings that need to be matched against Q. Algorithm F, however, will require only QSIZE accesses, one for each query list, if it is assumed that the table of SIZE values can be held in internal storage. Since a single disc access may take as much time as many thousands of similarity calculations, the advantages of this approach will become increasingly apparent as larger and larger files need to be processed. Accordingly it is suggested that algorithm F is the most efficient best match search procedure currently available: it forms the basis for the experimental best match structure search system described in Section 3.3.3.

5.4.2 Clustering experiments

Nearest neighbour searching is not only an important retrieval facility in its own right, but is also central to the efficient implementation of many of the clustering methods described in Chapters 2 and 4. This arises from the fact that many of them involve the assignment of compounds to the most similar cluster of compounds, as in the single pass or relocation methods, or the fusion of the most similar pair of clusters, as in the hierarchic agglomerative methods. The close relationship between nearest neighbour searching and clustering has been emphasised by several workers, in particular in a series of papers by Murtagh (1983, 1984a, 1984b), and it is thus of interest to determine the efficiency of algorithm F for clustering applications.

The most computationally demanding of the clustering methods described in previous chapters are those hierarchic methods which depend upon an NxN intermolecular similarity matrix which contains the similarities obtained from a comparison of each of the compounds in a data set with all of the other molecules. There is a similar computational requirement in the Jarvis-Patrick method which performed so well in the comparative clustering experiments of Chapter 4. For a collection of N compounds, at least $O(N^2)$ comparisons must be performed, and thus the use of such methods has generally been limited to the clustering of small data sets. As an example of this limitation, the CLUSTAN package, which was used for some of the experiments reported in Chapter 4, is restricted to data sets containing less than 1000 compounds, and most other generally available clustering packages are similarly constrained.

Although algorithm F was designed for the matching of a query structure against all of the compounds in a file of size N, it may be used with very little modification to calculate all of the N(N-1)/2 inter-molecular similarities as shown in algorithm G below.

```
FOR I := 1 TO N DO
  BEGIN
    FOR J := 1 TO I-1 DO
      V[J] := 0.0 ;
    FOR J := 1 TO M DO
      IF B[I,J] THEN
        FOR K := 1 TO I-1 DO
          IF B[K,J] THEN V[K] := V[K]+1 ;
    FOR J := 1 TO I-1 DO
      V[J] := 2*V[J]/(SIZE[I]+SIZE[J])
  END.
```

It may be noted in this pseudo-code that account need only be taken of the first I-1 molecules when calculating the similarities for the I'th molecule, owing to the symmetrical nature of the similarity matrix.

The efficiency of this procedure has been tested using a file of 7620 compounds taken from the Index Chemicus Registry System, with the structures being characterised by the augmented atom fragments contained within them. An implementation of the algorithm in Algol 68 required 60.8 minutes of CPU time on an ICL 1906S machine to evaluate the 29,028,390 distinct similarity coefficients; of these, just over 24 million were non-zero valued.

The actual amount of time used is, of course, dependent upon factors such as the machine, operating system and programming language used, and a better estimate of the efficiency of the algorithm may be made by comparing the running time with that required by a simple, brute-force comparison of all pairs of compounds, this being achieved by an appropriate modification of algorithm A. Timing constraints meant that it was not possible to process the entire file using this approach, but experiments with a systematic 1 in 5 sample suggest that the use of the brute-force algorithm would require a total of about 700 minutes of CPU time on the same machine. Thus, although the $O(N^2)$ complexity of the above modification of algorithm F is no less than that of a comparable modification of algorithm A, the constant of proportionality is sufficiently small in this implementation to allow an 11.5-fold speed-up (Willett, 1982).

5.5 ALGORITHMIC REFINEMENTS

It should be emphasised that the form of the algorithms presented above has been dictated, in part, by pedagogic considerations, and various optimisation strategies are possible. As an example, most of the coefficient calculations involve a term QSIZE in the denominator. This term may be ignored in a nearest neighbour search since the resulting coefficients will be monotonic with those obtained if it is included, and it is only the rank ordering of the coefficients that is of importance in determining the nearest neighbour, and not the absolute values; this might not be the case in certain clustering applications where the exact value is of importance. In addition, the discussion above has glossed over one or two points that need to be considered in the context of a practical implementation of the algorithms.

The first of these relates to the assumption that the molecules under consideration are characterised by binary fragment vectors, and that they can be represented for search by the bit map B. It has been noted elsewhere in this volume that it may be desirable, for some purposes, to use a weighting scheme based upon the occurrences of each fragment type in a structure. Such information can be taken into account, although with a certain degree of complexity in some cases, and the modifications that are required will be illustrated with reference to algorithms B and F.

Assume that B is an integer matrix, rather than a binary one, where B[I,J] contains the number of occurrences of the J'th fragment in the I'th molecule, and that the query is similarly described by an integer array, QA. Then algorithm F may be extended to allow for such weighting information merely by replacing

```
IF B[I,QA[J]] THEN
  V[K] := V[K] + 1
```

by

```
IF B[I[QA,[J]] > 0 THEN
  V[K] := V[K] + WQA[J]*B[I,QA[J]]
```

where WQA is an integer array, the J'th element of which, 1 <= J <= QSIZE, contains the frequency of occurrence in Q of the J'th of the QSIZE fragment types present. In addition, the calculation of the coefficient needs to be altered with QSIZE and SIZE[I] in the denominator being replaced by the sum of the elements in WQA and by the sum of the elements in B[I,] respectively. Efficiency would dictate that these sums be calculated just once when the file is created. Comparable sums need to be calculated if other coefficients, such as the cosine or Tanimoto coefficients, are to be used. In fact, if such an integer-based structural characterisation is to form the basis for the search system, the NxM matrix B might best be implemented as a sparse matrix, from which all of the zero-valued elements have been removed so as to reduce storage overheads.

In the case of the serial algorithm B, it is not sufficient to base the UPB calculation upon GSIZE[G] if weighting is used, since the structures will need to be grouped so that all of the compounds in a group have the same value for the sum of the weights in a row, B[I,],of the integer matrix B. Each such group is then characterised by a signature that contains the maximum frequency of occurrence in molecules within that group for each of the M fragments. Thus, if S[G,] is an M-element integer array which is the signature for the G'th block, containing NUM[G] compounds, it may be generated by

```
FOR I := 1 TO M DO
  S[G,I] := 0 ;
FOR I := 1 TO NUM[G] DO
  FOR J := 1 TO M DO
    IF B[I,J] > S[G,J] THEN S[G,J] := B[I,J].
```

The pre-processing and storage overheads of such an approach suggest that the algorithm might well be no more efficient in practice than the simple algorithm A.

The second point that needs to be considered is the exact form of the similarity coefficient used to determine the degree of similarity between Q and each of the structures in the file. The coefficient used in the examples is the Dice coefficient which, along with many other coefficients, has a zero value if there are no common fragments. Such structures may be removed by the use of the inverted file which restricts matching operations to those compounds that have at least one fragment in common with Q. In the case of distance measures, however, the nearest neighbour might possibly be a structure having no common fragments and the inverted file algorithms need to be modified to take account of this possibility.

Assume that the Hamming distance is being used as the similarity coefficient; then the true nearest neighbour may be obtained by the addition of the following code to the end of algorithm C:

```
FOR I := 1 TO N DO
  IF NOT INSPECTED[I] THEN
    IF QSIZE+SIZE[I] < BESTCOEFF THEN
      BEGIN
        BESTCOEFF := QSIZE+SIZE[I] ;
        NN := I
      END.
```

Note also that BESTCOEFF should be initialised to MAXINT and that the procedure NNUPDATE(I) should involve

```
IF COEFF < BESTCOEFF THEN
```

to reflect the fact that a dissimilarity coefficient is being used, rather than a similarity coefficient. Similar modifications are required for the other three inverted file algorithms reported here. It may, of course, be felt that a molecule should not be considered as a potential nearest neighbour if it has no features in common with Q; in this case, the modification may be ignored.

One area that has not been discussed so far is the use of <u>probabilistic</u> nearest neighbour searches (Weiss, 1981) in which near, rather than nearest, neighbours are acceptable as the outcome of a search. The basic idea here is that the upper bounds are exactly that, i.e. they represent the maximum possible value of the chosen similarity measure, whereas it is most unlikely that even the true nearest neighbour will have a similarity with Q which is anywhere near as large as the calculated upper bound. Probabilistic nearest neighbour searching involves the use of various heuristics to provide more realistic threshold similarities for the control of the extent of the search. Weiss (1981) and Eastman and Weiss (1982) discuss the use of such strategies for increasing the efficiency of multi-dimensional binary search trees, while Perry and Willett (1983) present a probabilistic version of algorithm D.

REFERENCES

Bentley, J.L., Weide, B.W. and Yao, A.C. (1980). Optimal expected-time algorithms for closest point problems. ACM Trans. Math. Soft. 6, 563-580.

Burke, J.M. and Rickman, J.T. (1973). Bitmaps and filters for attribute oriented searches. Int. J. Comput. Inform. Sci. 2, 187-200.

Burkhard, W.A. and Keller, R.M. (1973). Some approaches to best-match file searching. Comm. ACM 16, 230-236.

Eastman, C.M. and Weiss, S.F. (1982). Tree structures for high dimensionality nearest neighbour searching. Inform. Syst. 7, 115-122.

Friedman, J.H., Baskett, F. and Shustek, L. (1975). An algorithm for finding nearest neighbours. IEEE Trans. Comput. C-24, 1000-1006.

Friedman, J.H., Bentley, J.L. and Finkel, R.A. (1977). An algorithm for finding best matches in logarithmic expected time. ACM Trans. Math. Soft. 3, 209-226.

Fukunaga, K. and Narenda, P.M. (1975). A branch and bound algorithm for computing k-nearest neighbours. IEEE Trans. Comput. 24, 750-753.

Kittler, J. (1978). A method for determining k-nearest neighbours. Kybernetes 7, 313-315.

Lee, R.C.T., Chin, Y.H. and Chang, S.C. (1976).

Application of principal component analysis to multikey searching. ACM Trans. Soft. Eng. SE-2, 185-193.

Matthews, F.W. and Thomson, L. (1969). Weighted term search: a computer program for an inverted coordinate index on magnetic tape. J. Chem. Docum. 7, 49-56.

Murtagh, F. (1982). A very fast exact nearest neighbour algorithm for use in information retrieval. Inform. Technol. Res. Devel. 1, 275-283.

Murtagh, F. (1983). A survey of recent advances in hierarchic clustering. Comput. J. 26, 354-359.

Murtagh, F. (1984a). Complexities of hierarchic clustering algorithms: state of the art. Comput. Stat. Quart. 1, 101-113.

Murtagh, F. (1984b). A review of fast techniques for nearest neighbour searching, in COMPSTAT 1984. (International Association for Statistical Computing). Physica-Verlag, Vienna.

Naegeli, P.R. and Clerc, J.T. (1974). Computer system for structural identification of organic compounds from spectroscopic data. Anal. Chem. 45, 739A-744A.

Nevalainen, O. and Katajainen, J. (1982). Experiments with a closest point algorithm in Hamming space. Angew. Informat. 5, 277-281.

Noreault, T., Koll, M. and McGill, M.J. (1977). Automatic ranked output from Boolean searches in SIRE. J. Am. Soc. Inform. Sci. 28, 333-339.

Perry, S.A. and Willett, P. (1983). A review of the use of inverted files for best match searching in information retrieval systems. J. Inform. Sci. 6, 59-66.

Pfalz, J.L., Berman, W.J. and Cagley, E.M. (1980). Partial match retrieval using indexed descriptor files. Comm. ACM 23, 522-528.

Rasmussen, G.T., Isenhour, T.L. and Marshall, J.C. (1979). Mass spectral library searches using ion series data compression. J. Chem. Inform. Comput. Sci. 19, 98-104.

Shapiro, M. (1977). The choice of reference points in best-match file searching. Comm. ACM 20, 330-343.

Smeaton, A.F. and Van Rijsbergen, C.J. (1981). The nearest neighbour problem in information retrieval. An algorithm using upper bounds. ACM SIGIR Forum 16, 83-87.

Vallarino, O. (1976). On the use of bit maps for multiple key retrieval. ACM SIGPLAN Notices 11, 108-114.

Van Marlen, G. and Van den Hende, J.H. (1979). Search strategy and data compression for a retrieval system with binary-coded mass spectra. Anal. Chim. Acta 112, 143-150.

Weiss, S. F. (1981). A probabilistic algorithm for nearest neighbour searching, in Information Retrieval Research. (Edited by R.N. Oddy, S.E. Robertson, C.J. Van Rijsbergen and P.W. Williams). Butterworth, London.

Willett, P. (1981). A fast procedure for the calculation of similarity coefficients in automatic classification. Inform. Proc. Manag. 17, 53-60.

Willett, P. (1983). Some heuristics for nearest-neighbour searching in chemical structure files. J. Chem. Inform. Comput. Sci. 23, 22-25.

Willett, P. (1985). An algorithm for chemical superstructure searching. J. Chem. Inform. Comput. Sci. 25, 114-116.

Yuval, G. (1976). Finding nearest neighbours. Inform. Proc. Lett. 5, 63-65.

CHAPTER 6
Conclusions

The previous chapters of this book have described an extended series of experiments to evaluate the use of similarity and clustering techniques with machine-readable files of chemical molecules characterised by fragment substructures. This concluding chapter briefly lists the main conclusions which can be drawn from the experiments, and considers ways in which the work could be further extended in the future.

The first two chapters provide over-views of chemical information systems and similarity and clustering techniques, and thus provide the basis for the research work reported in the remainder of the book.

Chapter 3 considered the use of measures of inter-molecular structural similarity. It is clear from the property prediction experiments there that the measured structural similarities do indeed reflect similarities in molecular properties. It may be concluded, accordingly, that such measures provide a useful tool for identifying structures with activity characteristics comparable to those of a known compound. That said, it is by no means clear that the measures used are in any sense optimal for this purpose, and two other classes of similarity measure suggest themselves for comparative testing. The first of these are measures of the type considered by Randic and Wilkins (1979) and, in a more extended form, by Carhart et al. (1985). These measures are based upon the separations of atoms in molecules, and thus provide a more detailed representation of the overall topology of a molecule than do the atom-centred, fragment characterisations used in the work reported here. A still more detailed measure of the degree of similarity between a pair of structures is available from a consideration of the maximal common substructure (MCS), which corresponds to the maximal substructure common to a pair of graphs. Little attention has been given to the use of such graph matching techniques until quite recently, owing to the computational complexities of MCS identification. However, the increasing use of MCS algorithms for the indexing of chemical reaction data (Willett, 1986) means that there is now a fair degree of experience in the use of such algorithms with chemical graphs, and it seems not impossible that the algorithms could be adopted for similarity matching purposes. Thus the similarity measures considered in this book provide a fairly simple, low-level description of the similarity relationships between molecules; more complex measures might well prove yet more informative in elucidating the structural relationships between molecules.

Chapter 4 considered the application of a wide range of hierarchic and non-hierarchic clustering methods for the grouping of chemical structures. The systematic comparison which comprises much of this chapter is thought to be the first such study in the field of chemical information systems, and, within the constraints of the experimental environment, the results point strongly to the general utility of the Jarvis-Patrick clustering method; moreover, the effectiveness of this method is attested to by the promising, large-scale tests reported in Section 4.4. However, the evaluation of the hierarchic clustering methods was somewhat restricted since it was based only upon the bottom-level clusters, and it is possible that different results might be obtained if the hierarchy was to be analysed at a range of levels; this is a recurrent problem when evaluating hierarchic clustering methods (Mojena, 1977; Edelbrock, 1979). It would also be of interest to determine the extent to which the characteristics of the classifications are affected by the type of similarity measure which is used for clustering.

Chapter 5 presented efficient algorithms which may be used to increase the efficiency with which structural similarities can be calculated. In particular, an inverted file algorithm is described which would appear to be sufficiently fast in operation to allow best match searching in files containing tens or hundreds of thousands of structures, where a single query structure must be matched against each of the structures in the file. However, the algorithm is currently limited for clustering purposes to files containing only a few thousands of structures, owing to the much greater number of similarity calculations required when each molecule needs to be compared with all of the others. It is likely that the processing of substantially larger files will be achieved only with the advent of parallel computer hardware which would permit the distribution of the similarity computations over a number of different processors; examples of the potential of such hardware for information retrieval applications are described by Pogue and Willett (1984) and by Wipke and Rogers (1984).

In conclusion, it is important to note that the work

described in this volume is basic research, the practical utility of which can only be evaluated by extended testing of operational systems. A start has been made to the necessary developmental work with the systems described in Sections 3.3 and 4.4. However, many more such practical studies need to be carried out to determine precisely how similarity and clustering methods may best be used in chemical information systems.

References

Carhart, R.E., Smith, D.H. and Venkataraghavan, R. (1985). Atom pairs as molecular features in structure-activity studies: definition and application. J. Chem. Inform. Comput. Sci. 25, 64-73.

Edelbrock, C. (1979). Mixture model tests of hierarchical clustering algorithms: the problem of classifying everybody. Multivar. Behav. Res. 14, 367-384.

McGregor, J.J. (1982). Backtrack search algorithms and the maximal common subgraph problem. Software Pract. Exper.. 12, 23-34.

Mojena, R. (1977). Hierarchical grouping methods and stopping rules: an evaluation. Comput. J. 20, 359-363.

Pogue, C.A. and Willett, P. (1984). An evaluation of document retrieval from serial files using the ICL Distributed Array Processor. Online Rev. 8, 569-584.

Randic, M. and Wilkins, C.L. (1979). Graph theoretical approach to recognition of structural similarity in molecules. J. Chem. Inform. Comput. Sci. 19, 31-36.

Willett, P. (1986). Modern Approaches to Chemical Reaction Searching. Gower, Aldershot.

Wipke, W.T. and Rogers, D. (1984). Rapid subgraph search using parallelism. J. Chem. Inform. Comput. Sci. 24, 255-263.

Subject Index

Molecular graphics systems <u>see</u> Chemical graphics systems

Author Index